The Institute of Biology's
Studies in Biology no. 30

Cellular Radiobiology

by Christopher W. Lawrence
B.Sc., Ph.D.
Senior Scientific Officer,
Wantage Research Laboratory (A.E.R.E.)

Edward Arnold

First published 1971
by Edward Arnold (Publishers) Limited,
41 Maddox Street,
London, W1R 0AN

Boards edition ISBN: 0 7131 2335 4
Paper edition ISBN: 0 7131 2336 2

Printed in Great Britain by
William Clowes & Sons, Ltd, London, Beccles and Colchester

General Preface to the Series

It is no longer possible for one textbook to cover the whole field of Biology and to remain sufficiently up to date. At the same time students at school, and indeed those in their first year at universities, must be contemporary in their biological outlook and know where the most important developments are taking place.

The Biological Education Committee, set up jointly by the Royal Society and the Institute of Biology, is sponsoring, therefore, the production of a series of booklets dealing with limited biological topics in which recent progress has been most rapid and important.

A feature of the series is that the booklets indicate as clearly as possible the methods that have been employed in elucidating the problems with which they deal. Wherever appropriate there are suggestions for practical work for the student. To ensure that each booklet is kept up to date, comments and questions about the contents may be sent to the author or the Institute.

1971

INSTITUTE OF BIOLOGY
41 Queen's Gate
London, S.W.7

Preface

The rapid developments in the last few decades of the utilization of atomic energy and other applications of nuclear physics has brought into prominence the need for a better understanding of the action of ionizing radiations on living things. This is necessary not only to protect the present and future population against their harmful effects but also to enable the valuable properties of ionizing radiations to be efficiently and fully exploited. In addition, radiobiology is making a growing contribution to basic knowledge in the life sciences, and ionizing radiations have proved to be a useful tool with which to explore the structure and function of animals and plants. This booklet is intended to provide an introduction to some of the basic principles of radiobiology considered from this point of view.

Wantage, 1971 C.W.L.

Contents

Origins and Aims 1

1.1 The discovery of ionizing radiations and the beginning of radiobiology

Terrestrial life has evolved on a planet bathed in radiations of many different kinds which together are an important part of the environment to which all living things are adapted. Most of these radiations come from the sun, but others come from outer space and some from within the earth itself. Some, such as visible light, are essential for the very existence of life. Trapped by the photosynthetic mechanism of green plants, they provide directly or indirectly virtually all of the energy needed for the growth and multiplication of living things. Others, such as the ionizing radiations emitted by radioactive elements in the earth or the cosmic rays from space, are at best inessential and probably harmful; fortunately they occur at only very low intensities and living things are adapted to cope with their effects.

Logic would seem to dictate that *radiation biology*, usually shortened to *radiobiology*, should be concerned with the action of all these radiations on living things, but convention rules otherwise. The effects of ultraviolet and visible light fall into the province of *photobiology* while the term radiobiology is restricted to the study of the biological effects of the much more energetic ionizing, or atomic, radiations. This distinction is made because, as explained later, the two groups of radiations are absorbed in rather different ways.

Although men have always been exposed to very low levels of ionizing radiations, their existence was not appreciated until 1895. In that year Roentgen, while investigating the properties of electrical discharge through cathode ray tubes, discovered a new kind of radiation which he called X-rays. Tradition has it that he accidentally interposed his hand between the tube and a phosphorescent plate and saw a shadow of his bones, demonstrating the curious ability of these new rays to penetrate soft tissue. Shortly after this Becquerel discovered that somewhat similar radiations were emitted by uranium ores and Professor and Madame Curie isolated the radioactive element radium.

Becquerel is said to have 'burnt' himself by carrying some radium in his pocket, so that two properties of the new radiations were quickly discovered; their ability to penetrate living tissues and their ability to kill them. These properties were quickly applied to medical problems and after a remarkably short period X-rays were being used to examine the internal organs of living patients and the rays from radium to kill cancer cells in malignant tumours. Diagnostic and therapeutic *radiology*, as these two applications came to be called respectively, are widely practised today and ionizing radiations provide medicine with a powerful weapon.

The weapon is, however, two-edged though this was not immediately appreciated by the pioneer radiologists. In their efforts to establish the new techniques, they seem at first to have ignored warning symptoms such as dermatitis on the hand they used for testing their X-ray beam, by casting an image on a screen. Tragically, such dermatitis eventually developed into ulceration and then cancer, from which nearly all of the pioneer radiologists died. Such a development was difficult to foresee, since tumour induction took from ten to twenty years. An equally insidious hazard emerged when Muller, in the late twenties, discovered that ionizing radiations can produce gene mutations. Since nearly all mutations have detrimental effects, radiations pose a hazard not only to those exposed but also to their descendants in successive generations.

It therefore became imperative to estimate precisely what risk was entailed in exposure to different amounts and kinds of radiation, and to discover how best radiation could be used, both to minimize the harm from unnecessary exposure and to maximize the benefit when irradiation was necessary. This problem became all the more urgent following the rapid development of nuclear technology after World War II and the consequent greater use of radioactive materials in medicine, science, and industry, and much work has been devoted to it, leading today to the establishment of stringent regulations which govern and minimize our exposure to radiations.

Much of the original impetus for this work came, therefore, from medicine and radiological protection, and *radiobiology*, which is concerned more generally with the biological effects of ionizing radiations, only began to emerge as a distinctly separate field of inquiry when radioactive materials and other sources of radiation became more readily available. Today, radiobiology has grown into an extremely diversified field, including the interaction of radiations with ecosystems, populations, individual animals and plants, tissues, cells, and molecules. As in other areas of biology, many important problems in radiobiology are most suitably studied by observing effects in cells, rather than in tissues or multicellular organisms. Unicellular micro-organisms such as bacteria, algae, and fungi are often used for this work, and more recently artificial cultures of cells derived from the tissues of higher organisms, including man and other mammals. The basic aim of *Cellular Radiobiology* is therefore to understand how ionizing radiations affect cells, and at the same time to use this knowledge to find out how cells are constructed and the ways in which they carry out their normal activities.

1.2 Cellular radiobiology

One of the most remarkable properties of ionizing radiations, posing a problem central to cellular radiobiology, is their extraordinary effectiveness in disturbing the normal growth and development of cells. When cells are exposed to ionizing radiations a proportion of them eventually die or lose their ability to divide, some contain abnormal sets of chromosomes or transmit their chromosomes abnormally, while others exhibit heritable changes, that is, contain gene mutations. The proportion of cells affected in these ways usually rises with increasing exposure to the radiation, and such changes occur as a result of the cells absorbing some of the radiant energy. They are not uniquely produced by ionizing radiations of course and, in fact, many other agents like visible light or heat will also kill cells, damage their chromosomes, and so on. Ionizing radiations are unusually potent in this respect, however, and cells can be damaged by absorbing extraordinarily small amounts of their energy. The energy in a hot cup of tea would kill the drinker for example, if it were converted from heat to X-rays. The explanation for this potency lies partly in physics and partly in biology and some of the interesting aspects of radiobiology are concerned with attempts to relate these two fields.

The principles governing the ways in which ionizing radiations affect atoms during energy absorption have been known for many years, so radiobiologists have a well mapped starting point for their studies. An important feature of energy absorption is that ionizing radiations give up discrete and relatively large amounts of energy to only a few atoms within the irradiated material; the energy is not spread uniformly over all atoms and molecules as would be the case if the temperature of the material were raised. The amount of energy localized in this way is usually large enough to activate the molecule so that it readily takes part in chemical reactions, with the result that a more or less random selection of organic molecules within the cell suffer permanent chemical change. In many instances this chemical alteration will do little harm to the cell. Most kinds of molecules are replicated many times within each cell, so that loss of function in a few of them will be unimportant, while in other cases their function will be unimpaired by the chemical alteration, or the damage can be bypassed or repaired. In a few instances, however, irreparable damage will occur within certain critical structures, the integrity of which is essential to the cell, and serious metabolic disability will occur. The identity of all the critical sites, or 'targets' as they are sometimes called, is not known with certainty, but the very long threadlike molecules of deoxyribonucleic acid (DNA), the genetic material, are very important targets and possibly the only important targets except when the effects of very high doses are considered. Chemical changes in DNA or other molecules usually occur very quickly after energy absorption but the ultimate consequences of this damage are not immediately manifest, and may take hours, days, or even years to develop. During this period the cell's metabolism becomes increasingly disturbed and the molecular damage is 'amplified' to give rise to easily discernible changes in cells.

Exposure of cells to ionizing radiations therefore sets off a chain or network of reactions giving rise first to chemical and then to metabolic or

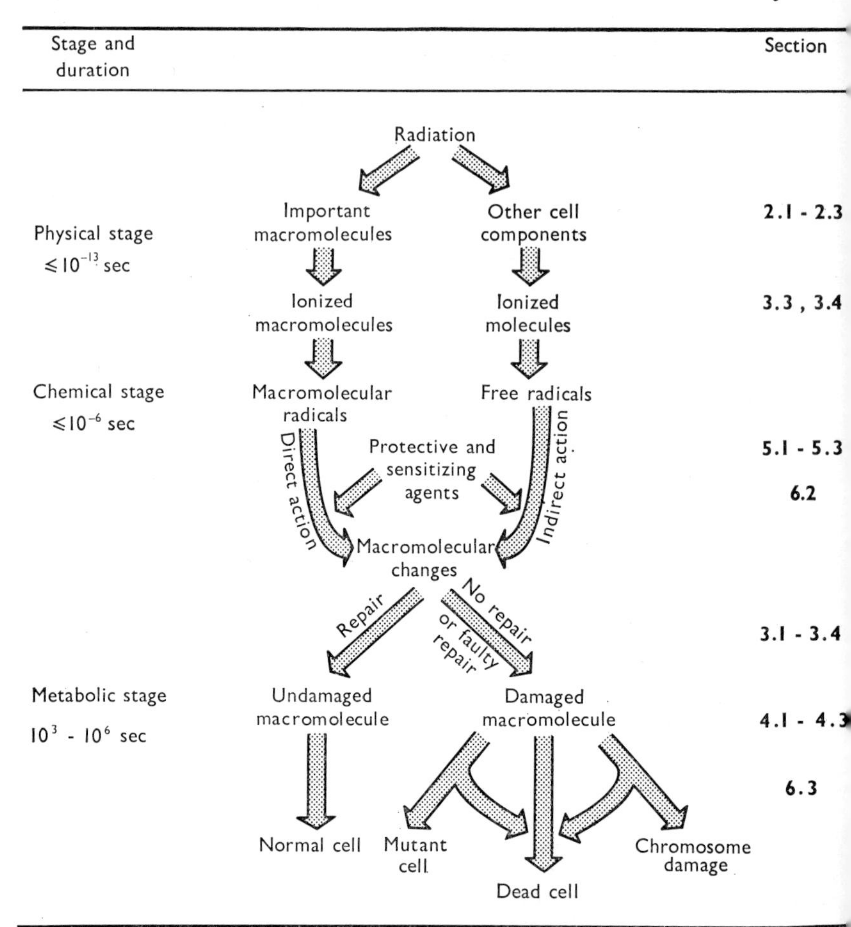

Fig. 1–1 General scheme for the development of radiation damage in cells.

physiological changes (Fig. 1–1), and it is the long term aim of cellular radiobiology to uncover all the ramifications of this sequence of events. At present, some parts of the sequence are known much better than others and it is usually difficult to relate observations in one part to those at another. Overall, however, three main stages can be recognized (Fig. 1–1). The *physical stage* is concerned with energy absorption processes, which occur extremely quickly (10^{-13} s) and over dimensions measured in atomic diameters. Although the principles of energy absorption were established long ago, the details are only now becoming clear, especially when cells are involved. The *chemical stage* covers the period in which activated molecules react with one another and with the normal cell constituents, and ends when chemical stability is restored. This usually takes place within a millionth of a second and involves distances of the order of a nanometre or less. Important problems at this stage concern the identity of the targets, the kinds of damage which inactivate them, and the ways in which this damage is produced. Finally the *physiological stage* is concerned with the metabolic consequences of the radiation-induced biochemical change, and may take more than 10^6 s to complete. Important problems at this stage are the ways in which biochemical damage is translated into metabolic disturbance, and the mechanisms by which cells can repair or bypass damage.

The Nature of Ionizing Radiations and their Interactions with Matter — 2

2.1 Types of ionizing radiation

Radiations are physical phenomena in which energy travels through space, without the aid of a material medium. *Ionizing* radiations include a variety of highly energetic radiations having in common the ability to eject electrons from atoms in the matter through which they pass. This leads to the production of positively and negatively charged *ions*, hence the name ionizing radiations. Radiations of this type fall into one or other of two classes, *corpuscular* (or particulate) and *electromagnetic* radiations.

Corpuscular radiations consist of streams of atomic or sub-atomic particles moving at high velocities and which therefore carry kinetic energy; that is, the particles' energy is determined by their mass and velocity, though some particles may also contain energy in their internal structure. Table 1 lists the particles concerned with some of the commonly

Table 1

Name of particle	*Approximate mass* (amu)	*Charge*
β-particle (electron)	5.5×10^{-4}	-1
Proton	1.0	$+1$
Neutron	1.0	0
Deuteron	2.0	$+1$
α-particle (nucleus of helium atom)	4.0	$+2$

encountered corpuscular radiations. The electric charge of these particles is expressed in multiples of that of the electron, which possesses the smallest amount of free electric charge known to exist, while their mass is given in atomic mass units (amu), defined as one sixteenth of the mass of one atom of the common form or isotope of oxygen, approximately equal to 1.7×10^{-24} g.

The velocity of the particles in ionizing radiations is generally high, especially with the lighter particles such as electrons which may approach, but not exceed, the velocity of light. Their kinetic energy is not limited, however, since attempts to increase the velocity of particles already travelling at appreciable fractions of the speed of light lead principally to an increase in mass, which can increase without limit. It is customary to express the kinetic energy of particles in electron volts (eV), one eV being equal to about 1.6×10^{-19} J. Corpuscular radiations with energies above a few hundred eV are all capable of producing ionizations, but the radiations commonly used in radiobiology usually lie between a few keV (1 keV $= 10^3$ eV) and about 10 MeV (1 MeV $= 10^6$ eV).

Electromagnetic radiations include radiowaves, infra-red, visible light, X-, and gamma-waves, though only the last two are ionizing radiations. They consist of vibrating electromagnetic fields which travel through space

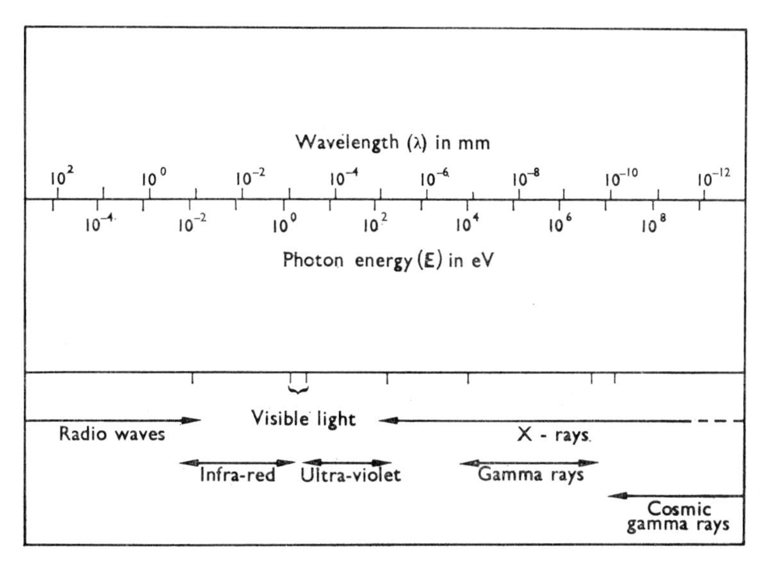

Fig. 2–1 Wavelengths and photon energies of electromagnetic radiations.

at a velocity of nearly 3×10^8 m/s, the frequency of vibration, wavelength, and velocity of propagation being related according to the equation

$$\nu = \lambda c$$

where ν is the frequency, λ the wavelength, and c the velocity of propagation. The wavelengths of various electromagnetic radiations are given in Fig. 2–1. Although these radiations consist of continuously distributed waves, when they are involved in energy interchanges with matter they are found to be composed of discrete packages of energy called photons or quanta. Different radiations have photons possessing characteristic amounts of energy, the amount being given by

$$E = h\nu$$

where E is the energy in joules, h is Planck's constant, 6.6×10^{-34} J/s, and ν is the number of vibrations per second as before. It follows from the two equations above that radiations can be described by either the frequency of the vibrations, the wavelength, or the photon energy in an entirely interchangeable way. For radiobiological purposes photon energy is generally the most useful and as with the kinetic energy of corpuscular radiations, it is usually expressed in electron volt units. Electromagnetic radiations sufficiently energetic to produce ionizations are called X-rays if produced by machines or gamma rays if emitted by radioactive elements, but apart from their origin there is no essential difference between X- and γ-rays.

2.2 Interactions between radiations and matter

When ionizing radiations pass through matter they give up some or all of their energy to the surrounding atoms and molecules, that is the matter absorbs their energy. The processes by which this energy absorption occurs vary according to the energy of the radiation concerned and whether it is composed of photons, charged or uncharged particles, but in each case the energy-absorbing event depends on an encounter, or *interaction*, between a photon or particle of the radiation and a sub-atomic particle, either electron or atomic nucleus, of the absorbing material. When charged particles pass through matter, for example, they experience forces of attraction and repulsion owing to the effect of their electric charge on that of the surrounding electrons and nuclei. Beta-particles are attracted by the positively charged nucleus and repelled by the negatively charged orbital electrons, while α-particles behave in the opposite way.

These forces tend to slow down and deflect the particle from its original path, the light or slow particles being deflected more strongly than the heavy or fast ones. The atoms and particles of the matter will also be disturbed by these encounters, which are called collisions though they are not collisions in the ordinary sense of the word since they can occur even though the particle comes no closer to the atom than 500 atomic diameters, owing to the spread of their electric fields. When the particles move relatively slowly, the surrounding atoms are merely displaced a little by these collisions and quickly return to their former positions. When the fast and energetic particles characteristic of ionizing radiations travel through matter, however, the atoms in their path have insufficient time to accommodate themselves to the presence of the particles and as a consequence of these *inelastic collisions* become highly disturbed, oscillating and vibrating violently. In some atoms an electron may be so disturbed as to jump out of its normal orbit either into another orbit further from the nucleus or, if it receives enough energy from the incident particle, even to move right away from the atom. Atoms with electrons in orbits other than the usual ones are said to be *excited* and are more chemically reactive than usual, while atoms which lose electrons become positively charged ions which, as well as the free electrons, are extremely chemically reactive. Interactions between particles and electrons of this kind are by far the most common energy-absorption events in most radiobiological circumstances, and ionizations are probably the most significant events in the production of biological damage. Other kinds of interactions, such as those between particles and nuclei, are much less frequent.

Most incident particles are sufficiently energetic to impart more energy to the electron than is required merely to eject it from the atom, and the ejected electron may acquire a high velocity, that is gain kinetic energy, and may then itself give rise to further ionizations and excitations. The electrons liberated in these ionizations may in turn produce further ionizations and so on, though with rapidly decreasing probability. Usually the ejected electron is capable of producing only a few ionizations and these occur close to the original event to form an 'ion cluster'. Some, however, have a greater range and may form a δ-ray, which is often arbitrarily defined as an electron capable of producing thirty or more ionizations. As the incident particle travels further through matter, it is involved in an increasing number of collisions, each taking a little of the particle's energy. The particle therefore slows down and eventually will travel too slowly, that is will possess too little energy, to produce ionizations or even excitations, and its remaining energy is dissipated as heat. If the incident particle is an electron it will be captured by an atom. If it is an α-particle, it will capture two electrons to form a helium atom.

The original kinetic energy of the particle is therefore degraded into smaller and smaller pieces and these are spread out, despite the essential randomness of collisions, in a way which is more or less a characteristic of the type of particle and its original energy. The path of the particle forms a track of ion clusters from which δ-rays radiate in all directions, the average distance between ion clusters depending on the velocity and charge of the particle. This distance will be relatively great for fast electrons but some thousands of times smaller for α-particles of the same energy, because α-particles are much more massive than electrons and therefore their velocity is much lower at any given level of kinetic energy. They also possess twice the electric charge of the electron and therefore interact more strongly with the surrounding atoms. Differences between radiations depending on the spacing of ion clusters are usually expressed in terms of 'linear energy transfer' (LET), defined as the energy deposition per unit length of the track of the original particle and of the electrons it releases.

The only uncharged particles commonly encountered in radiobiology are neutrons and these, lacking charge, cannot directly produce ionizations in the same manner as charged particles. Fast neutrons, in the energy range 10 keV to 10 MeV, lose most of their energy by elastic collisions with the nuclei of light elements, particularly hydrogen, because the hydrogen nucleus contains a single proton, which has a mass almost identical to that of the neutron. Elastic collisions between neutrons and much lighter particles such as electrons, or much heavier ones such as the nuclei of heavy elements, are much less probable and in hydrogen-rich material like living tissue, 90 per cent of the energy lost by fast neutrons involves collisions with hydrogen nuclei. During such collisions the protons acquire a velocity which varies up to that of the incident neutron and they then produce ionizations and excitations in the manner described previously. Slow neutrons, with energies less than 100 eV, such as those slowed down by collisions with protons, may be captured by a nucleus which then emits a γ-ray photon. These γ-rays can also give rise to ionizations.

Unlike charged particles, photons can pass through matter without trace, but occasionally a photon will interact with an orbital electron or atomic nucleus, the probability of these events depending on the photon energy (Fig. 2–2). Within the range of photon energies normally associated with

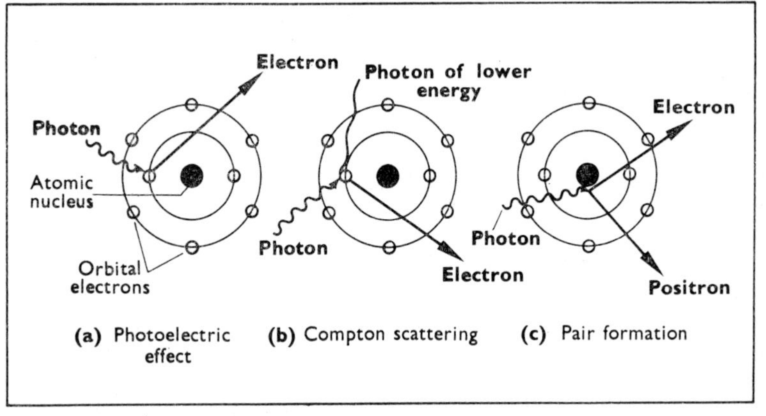

Fig. 2–2 Interactions between X- or γ-ray photons and matter.

X- and γ-rays only three kinds of interaction are common. At low energies *photoelectric absorption* predominates, in which a photon gives up all of its energy to an orbital electron which is then ejected from the atom. Usually only a small fraction of the photon's energy is required to eject the electron, the remainder being carried off by the electron as kinetic energy. This high velocity electron will then give rise to ionizations and excitations as described earlier. At intermediate photon energies, *Compton scattering* becomes the most probable interaction. In Compton scattering a photon gives up only a portion of its energy to an orbital electron, which is again ejected from the atom at high velocity. The photon, now of lower energy, is deflected from its previous path and may become involved in further interactions. Finally, high energy photons can interact with atomic nuclei to create an electron and a positron, any excess energy remaining after the creation of these particles again being carried off as kinetic energy. This kind of event is called *pair production*. Apart from the ionizations and excitations which these particles can produce, the positron will, when most of its kinetic energy is lost, combine with and annihilate an electron, producing two γ-rays in the process.

Despite the variety of ways in which different ionizing radiations deposit energy, there are three very important features common to them all. First, all the energy-absorption events depend on interactions between the photons or particles of the radiation and the sub-atomic particles, such as nuclei and electrons, of the absorbing matter; the way in which these particles associate to form atoms and molecules is relatively unimportant. This feature is unique to ionizing radiations and is not true for others, such as ultra-violet radiation or visible light. Visible light, for example, is absorbed strongly by some molecules but hardly at all by others with similar or identical atomic composition; that is, absorption of light depends strongly on how atoms associate. With ionizing radiations, however, only the atomic composition of the absorbing material is important. In fact, with the exception of uncharged particle radiations, the majority of absorption events involve interactions between the radiation and orbital electrons, so that the amount of energy absorbed is approximately proportional to the electron density of the material.

Second, as their name implies, a feature shared by all ionizing radiations is their ability to eject orbital electrons from some of the atoms of the material through which they pass, and therefore produce positive and negative ions. These chemically active ions are probably responsible for most of the significant radiobiological damage. Moreover, in most cases the ejected electron distributes the majority of the absorbed energy, and so represents another feature common to all ionizing radiations. It is also important to emphasize that except at very high doses only a small fraction of atoms become ionized, and the amount of energy required to produce the ions is large compared with that required for most chemical reactions; that is, energy is deposited in discrete and fairly large units located within only a few molecules of the irradiated material. It is not spread out over all molecules in the material as would be the case if its temperature was raised.

Third, interactions between ionizing radiations and atomic particles occur essentially at *random*, and it is not possible to predict whether any particular electron or nucleus will respond to the radiation, or in what way it will respond. When living things are exposed to radiation, therefore, an approximately *random* selection of atoms will be ionized, and different cell components will absorb energy roughly according to their proportion by weight.

The various radiations differ chiefly in their ability to penetrate into matter and in the spatial distribution of the ionizations and excitations that they produce. Electromagnetic radiations penetrate deeply into matter, which merely attenuates rather than completely stops them. Uncharged particles also penetrate deeply, but charged particles have a definite range, the size of which depends on the type of particle, its energy, and the kind of material through which it passes. The charged particle radiations used in radiobiology rarely penetrate more than a few centimetres into living tissues, and often much less. Such particles also transfer their energy to a large number of orbital electrons, each of which therefore receives a relatively small proportion of the total. As a consequence, even if these secondary electrons are capable of producing further ionizations, such events are usually clustered close together and close to the path of the incident particle. An X- or γ-ray photon, on the other hand, transfers a large proportion of its energy to the electron or electrons with which it interacts, and these usually acquire a high velocity. The ion clusters are therefore more widely spaced, except in the δ-rays and at the ends of the electron paths.

2.3 Units of radiation dose

It is of course essential to measure the size of the radiation dose in radiobiological experiments, and while radiation dosimetry is beyond the scope of this short book, the problem of what units are to be used must be mentioned briefly. Ideally, radiation dose should be expressed as the number of ion pairs and excited atoms in a given mass of living matter, but unfortunately it is impossible to count these directly. Instead it is usual to measure the amount of energy absorbed by a given mass, to give the *absorbed dose*. The unit of absorbed dose is the *rad*, which is that amount of radiation which deposits 10^{-2} J/kg. Although energy absorption is virtually independent of the way sub-atomic particles are aggregated into atoms and molecules, it is not independent of atomic composition. Strictly speaking therefore, measurements of absorbed dose should be made in the particular experimental material irradiated. Fortunately, living things are composed largely of light elements, and except at very low and very high energies in the ionizing range, variations in atomic compositions are of negligible importance so that in most instances living matter can be regarded, for the purposes of dosimetry, as being composed entirely of water. This is not true for neutron radiations, however, where atomic composition strongly influences energy absorption and therefore must always be determined for each material irradiated.

Apart from measuring absorbed dose, it is often useful to measure the output of an X-ray machine or source of γ-rays. The *exposure dose* is defined in terms of a standard absorbing material, air being chosen for this purpose. The unit of exposure dose is the *roentgen* (r) being the quantity of X- or γ-radiation such that the associated corpuscular emission (i.e. electrons) per 0.001293 g of air produces, in air, ions carrying 1 electrostatic unit of electricity of either sign (0.001293 g of air occupies one cubic centimetre at 0°C and 760 mm pressure). Conveniently, exposure of many biological materials to a dose of one roentgen results in energy absorption of approximately one rad.

The Lethal Effects of Ionizing Radiations 3

3.1 The nature and definition of cell death

Despite the enormous diversity of shapes and sizes that the cells of animals, plants and bacteria can assume and the variety of their activities, they are all constructed from similar materials and on broadly similar lines featuring, for example, genetic material composed of deoxyribonucleic acid (DNA), enzymes made of protein, limiting and internal membranes, and so on. Such morphological and biochemical features constitute the static aspects of cells. From the dynamic point of view these cellular components are organized into an intricate and delicately balanced system, adapted during the course of evolution to maintain itself in the face of random internal changes and variation in the extra-cellular environment. Many cells also actively attempt to maintain continued growth and multiplication, again in the face of factors which tend to oppose or divert them from such activities.

When cells are exposed to ionizing radiations, a more or less random selection of their atoms engage in chemical reactions, so that some of their constituent molecules or structures sustain loss or change of function. Radiation therefore presents an additional stress to the cells which also tends to disturb their organization. The extent of this disturbance will depend partly on the amount of chemical change, which in turn will depend on the absorbed dose, but much more importantly on the kinds of molecules damaged and the kinds of damage they receive. As a consequence of this biochemical damage, a great variety of changes can be observed in irradiated cells, some temporary and others permanent. Cells capable of dividing, for example, often temporarily 'mark time' after irradiation but eventually resume their normal cycle of growth and division. In other cases the changes are more permanent and lead for instance to the death of the cell. A variety of symptoms can therefore be used to evaluate radiation damage in cell populations. Among these, the ability to survive is perhaps the broadest or most general measure of cellular activity, and consequently has been widely used as an indicator of radiation damage.

It is important to define what exactly is meant by cell 'death', since different definitions and criteria can be used. Most commonly a cell is defined as 'dead' if it fails to produce by repeated division a group or *colony* of more than fifty descendant cells. This is more accurately called loss of *proliferative capacity*, and the choice of the number fifty is purely an arbitrary one. Irradiated cells must lose their proliferative capacity for many different reasons, and do so in several ways. Some disintegrate or *lyse*, destroyed by their own enzymes, while others merely fail to divide although they continue to grow and form 'giant' cells or long filaments as in the case of bacteria. These are dead only in a reproductive or genetic sense. Lysis and 'giant' formation may occur directly in the cells which have been irradiated, or in their daughters of the next or subsequent cell generations, though the proportion doing so rapidly decreases in succeeding generations after irradiation. Cells which lose their proliferative capacity usually divide at least once after a dose which allows many of them to survive, but after higher doses 'death' usually occurs without an intervening division.

3.2 The identity of the 'targets'

It is of course important to attempt to identify the molecules, structures, or 'targets' whose damage is eventually responsible for loss of proliferative capacity. Several lines of evidence clearly show that the most important of these are located within the cell nucleus. First, experiments in which irradiation is confined to only part of each cell show that much larger doses are required to kill them if the nucleus is excluded from the irradiated portion. ULRICH (1951), for example, found that several hundred times the dose of X-rays was needed to kill *Drosophila* eggs if the nucleus was shielded from the radiation. Similarly, VON BORSTEL and ROGERS (1958), using a microbeam of α-particles, found that eggs of the wasp *Habrobracon* could be killed by the passage of a single particle through the nucleus, although sixteen million particles through the cytoplasm were needed to kill half of the eggs. The same conclusion was reached by COLE (1965), who irradiated cells with electrons of varying energy which penetrated to varying depths into the cell. Those which could penetrate as far as the nucleus were far more lethal than those which could not.

Second, the work of SPARROW (1965) suggests not only that the targets are located within the nucleus but also that chromosomes are the sites of lethal damage. There is an enormous variation in radiosensitivity between different species of higher plants, more than a hundred times the dose being required to kill half of a sample of plants of one species compared with another. Sparrow found that this variation was highly correlated with variation in the average amount of DNA per chromosome. Radiosensitive species possess a few large chromosomes, each containing a large amount of DNA, while radioresistant species possess many small chromosomes, each containing only a small amount of DNA. Putting this in a different way, the average amount of energy absorbed within the DNA of a single chromosome is almost constant with respect to any given level of damage in all higher plants.

Finally, ERIKSON and SZYBALSKI (1963) have shown that cells grown in the presence of halogen-substituted pyrimidines will incorporate these analogues into their newly synthesized DNA in place of some of the normal pyrimidine bases. Such cells become more sensitive to the lethal effects of radiation and, moreover, the degree of sensitization is proportional to the amount of analogue incorporated, again pointing to a DNA target.

The conclusion that loss of proliferative capacity is usually caused by damage within DNA molecules is also supported by arguments based on general principles. DNA represents the 'blueprint' for the construction and maintenance of the cell and as genetic material performs the function of storing, replicating, and transmitting the information or set of instructions necessary for the synthesis and regulation of enzymes, hence controlling all aspects of metabolism, structure, and development. Clearly, irreparable damage to DNA, errors in its replication or its faulty transmission to daughter cells could have serious biological consequences which might easily lead to the death of the cell or its daughters. In addition, DNA forms very large *macromolecules*, with molecular weights up to 10^9 or more, and large molecules are subject to more damage than an equal number of small ones. Moreover, there is probably only a single copy, or at most only a few copies, per cell of each DNA molecule, so that their damage is likely to have serious consequences. BACQ and ALEXANDER (1961) have calculated that a dose of one hundred rads produces only about three hundred chemical reactions in organic molecules per cubic micron of tissue, so that the number of changed molecules within cells must be a very small proportion of the total. This dose is nevertheless capable of killing quite a high proportion of cells, particularly of the more radiosensitive species. Such damage is distributed among different kinds of organic substances within the cell approximately according to their proportion by weight. That is, if the material in question forms one per cent by weight of such substances, it will undergo only about three chemical reactions in each cubic micron. Since there are usually tens or hundreds of copies of most kinds of organic molecule, such as a given enzyme, in each cell only a tiny fraction of them will be damaged and loss of function in these is unlikely to be very important. The components potentially most at risk within irradiated cells are therefore the unique macromolecules and of these the long threads of DNA must be the most important. Nevertheless it cannot be concluded that *all* cells die because of damage to their DNA, and some cell death due to damage in other macromolecules such as proteins cannot be ruled out.

3·3 Dose–survival curves

Dose–survival curves, obtained by plotting on a graph the proportion of cells which retain their proliferative capacity against dose, provide a useful way of assessing and investigating radiosensitivity in populations of cells. Although such curves vary according to the particular cells used and the conditions under which they are grown both before and after irradiation, most tend to assume one or other of two basic shapes. Curve 1 of Fig. 3–1 is characteristic of survival curves of bacteria and some haploid cells grown in certain conditions, while curve 2 is commonly found with diploid cells. Curve 1 is also found when some haploid and diploid cells are exposed to high LET radiations (see p. 9), and is typical of inactivation curves of viruses and enzymes. Historically, the problems posed by the shapes of these curves led people such as Dessauer, Crowther, Lea, Zimmer and others (see LEA, 1946; ZIMMER, 1961) to develop the 'Hit' and 'Target' theories, more properly called hypotheses, which provided the foundation for many of the subsequent developments in cellular radiobiology. Before

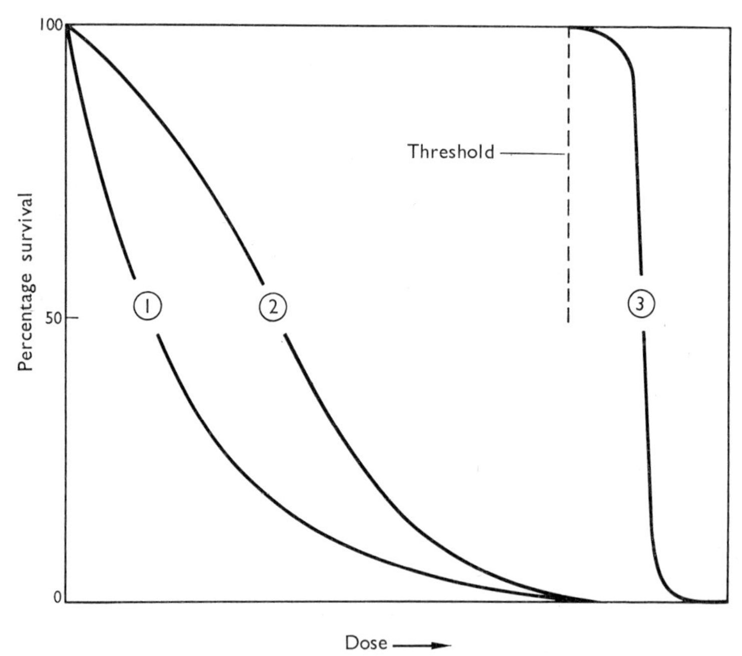

Fig. 3–1 Characteristic dose–response curves for loss of proliferative capacity.

these ideas emerged, radiobiologists were puzzled by the contrast between the dose-survival curves characteristic of irradiated cells (Fig. 3–1, curves 1 and 2) and those which they believed to be typical for the action of poisons and other chemical agents (Fig. 3–1, curve 3). Low doses of poisons were thought to be virtually ineffective up to a threshold dose, but even very small doses of radiation killed at least some cells. 'Hit' theory attempted to explain this in terms of the basic processes of energy absorption in the following way.

Although it is not obvious when large volumes of tissues are considered, energy absorption is *quantized*, that is, involves discrete units of energy, and this is most important when very small volumes such as that occupied by a DNA molecule are examined. These units are of course the amount of energy required to ionize or excite an atom, etc., and their existence means that when a population of cells is exposed to radiation, the targets they contain will absorb energy in a quantized manner. If the particular absorption event that inactivates a target is called a 'hit', individual targets will receive whole numbers of hits, either 0, 1, 2 or n of them. For the present purpose it is not necessary to define exactly what a 'hit' is, nor yet what the target

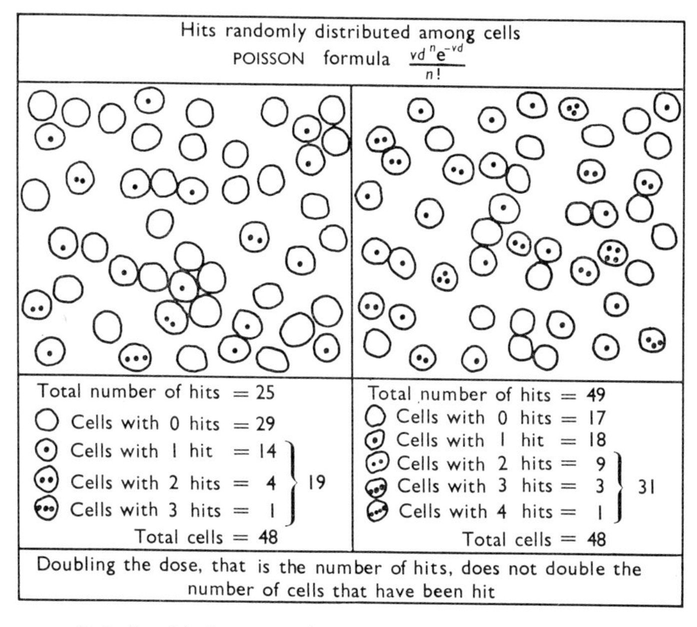

Fig. 3–2 Relationship between the average number of hits and the number of targets that have been hit at least once. Doubling the average number of hits does not double the number of hit targets.

is. If a population of cells, each containing an identical target with a volume v, is exposed to a dose of radiation d, expressed in 'hits' per unit volume, the average number of hits per target will be vd. By definition, the average number of hits per target is strictly proportional to dose, so that doubling the dose doubles the average number of hits, but it does not double the number of *targets that have been hit at least once* because at higher doses there will be an increasing proportion of targets that have been hit two or more times (Fig. 3–2). Since interaction between radiation and atoms is essentially a random process, the hits will be distributed randomly among the targets and in this circumstance the proportion of targets receiving exactly n hits, where n can be any whole number, is given by Poisson's formula,

$$\frac{(vd)^n e^{-vd}}{n!}$$

where e is the base of natural logarithms, approximately equal to 2.7, and $n!$ stands for factorial n.

$$n! = n \times (n-1) \times (n-2) \ldots 3 \times 2 \times 1$$

If a single hit is sufficient to inactivate a target, and therefore kill the cell, those with more than one hit will have been overkilled and only cells in which the target received no hits will retain their proliferative capacity. The proportion of such cells can be obtained by putting $n=0$ in the above formula, which gives

$$\text{surviving fraction} = e^{-vd} \text{ (one hit curve)}$$

That is, the surviving fraction is exponentially related to dose, giving curve 1 in Fig. 3–1. If two hits are needed to inactivate the target, the surviving fraction will be augmented by a proportion given by the second term of the Poisson expansion (putting $n=1$) and so on for others in the multihit family of curves, which have the same general shape as curve 2 in Fig. 3–1. A similar family of curves can also be generated by assuming that cells contain several targets, each of which must be independently inactivated to kill the cell. If there are two targets, both of which must be hit at least once to kill the cell, the proportion of cells in which the first of the targets *is* hit at least once is $1-e^{-vd}$ and of these a fraction $1-e^{-vd}$ will also receive hits in the second target. The proportion of cells in which both targets are inactivated will therefore be $(1-e^{-vd})^2$ and the proportion of survivors is one less this fraction,

$$\text{surviving fraction} = 1-(1-e^{-vd})^2$$

If there are m targets each of which must be hit at least once,

$$\text{surviving fraction} = 1-(1-e^{-vd})^m \text{ (multitarget curve)}$$

Since all of these curves contain exponential components, it is convenient to plot the *logarithm* of the surviving fraction against dose, as in Fig. 3–3. For the one-hit curve this gives a simple linear relationship with dose (Fig. 3–3 curve 1),

$$\log \text{ (surviving fraction)} = -vd$$

while multihit and multitarget curves become approximately linear after an initial 'shoulder' in the curve at low doses (Fig 3–3 curve 2).

Hit theory therefore shows that the general features of survival curves are a natural consequence of the existence of targets and of the random

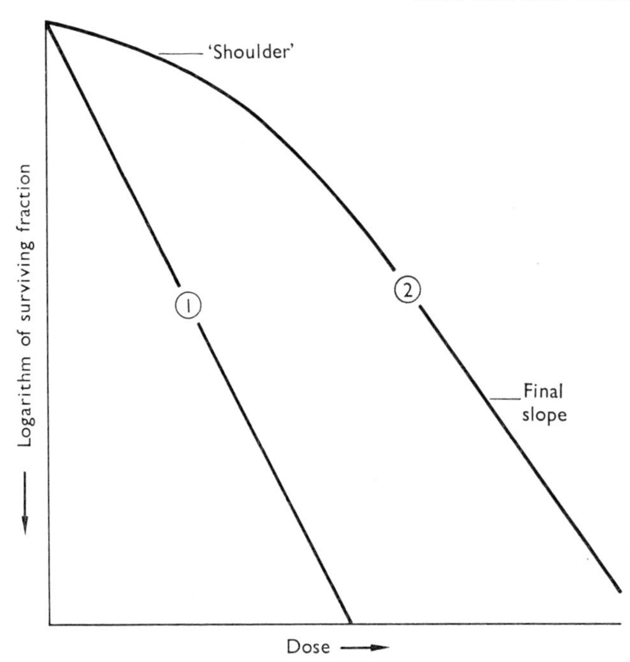

Fig. 3–3 Survival curves plotted on a semi-logarithmic scale.

and quantized nature of energy absorption. The predicted exponential relationship, with or without an initial 'shoulder', between surviving fraction and dose has been demonstrated frequently, and some of the best experimental results are shown in Fig. 3–4. This shows that loss of proliferative capacity in the bacterium *Serratia marcescens* is exponentially related to X-ray dose over nearly ten powers of ten, that is up to doses where only one bacterium in ten thousand million gives rise to a colony. This is in remarkable agreement with theory and represents a great technical achievement.

Despite this general agreement between theory and observation, it is now clear that past attempts to estimate the number of targets per cell and the number of hits required to inactivate them by finding the theoretical curve which best fitted the data were invalid. This is not only because different assumptions may lead to very similar curves or because cells are rarely identical in radiosensitivity, but also because the models used were too simple. It is now known that cells can repair some of their damage, especially after low doses, and repair as well as the existence of several targets or the requirement for several hits can give rise to a 'shoulder' on survival curves. Similarly the use of *Target theory* to calculate target volumes must now be restricted to a few special cases. According to the simple multihit and multitarget models given above, if dose is expressed

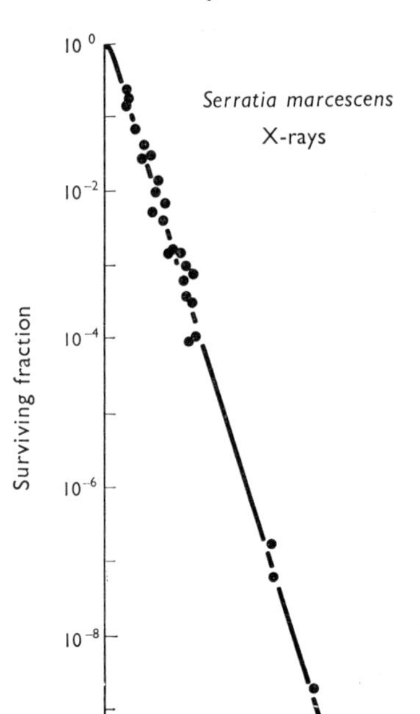

Fig. 3–4 Survival curve for the bacterium *Serratia marcescens*. (Courtesy D. L. Dewey, *Radiation Research*, **19**, 64–87, and Academic Press)

in hits per unit volume it should be possible to calculate v, the target volume, from the slopes of survival curves. As shown by the work of Sparrow (see § 3.2), these are steep in species with large targets and less steep in those with small targets. Volumes calculated in this way are likely to be very inaccurate, however, not only for the reasons given above but also because it is rarely possible to say exactly what constitutes a hit. Target theory has been used with greatest success to calculate the molecular weights of enzymes and the nucleic acid portion of viruses.

3·4 The influence of radiation quality

Although different kinds of ionizing radiations all interact with matter in a basically similar manner, they vary with regard to the spatial distribution of ionizations and other energy absorption events within the irradiated material. Consecutive ion clusters are well separated along the track of the incident charged particle or ejected electron with *sparsely* ionizing radiations, like γ-rays or very fast electrons, but are much closer together with *densely ionizing* radiations, like protons or α-rays (Fig. 3–5). These differences in radiation quality are expressed in terms of linear energy transfer (LET), the average rate of energy deposition per micron of track of the incident particle or ejected electron (see 2.2). Sparsely ionizing radiations have low, and densely ionizing radiations high, LET.

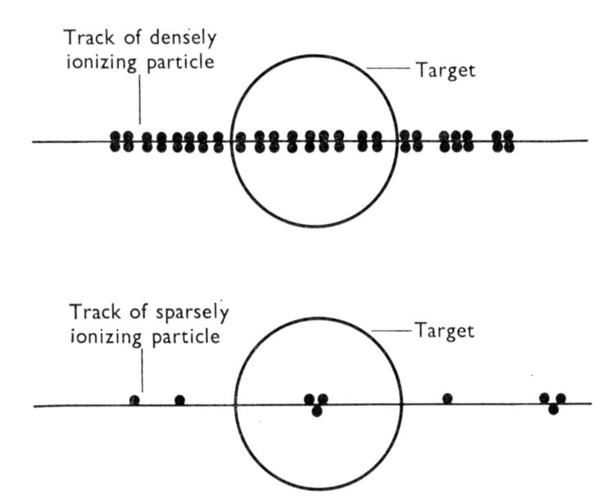

Fig. 3–5 The number of ionizations occurring within targets that are hit by radiations of different quality. Densely ionizing radiations produce many ionizations within targets, sparsely ionizing radiations only a few.

Radiations with different LET's kill cells with varying efficiency, some requiring a greater dose to kill a certain proportion of cells than others, and the ratio of the effectiveness of a given radiation to that of some standard, often the γ-rays from an isotope of cobalt, is called its *relative biological effectiveness* (RBE). Results from several experiments show that the relationship between RBE for loss of proliferative capacity and LET varies according to the particular type of organism studied (Fig. 3–6). The RBE for virus inactivation decreases steadily with increasing LET and a similar picture, at least at high LET values is found with bacteria. The RBE for survival in mammalian cells, on the other hand increases to a maximum and then drops off, and this is typical of all cellular organisms except bacteria. In mammalian and other similar cells, increase in RBE depends both on decrease in 'shoulder', where present, and increase in the slope of the exponential region of the survival curve.

Three kinds of explanation for these results have been proposed. According to the multihit explanation, viruses are inactivated by a single hit, bacteria by a few, and other cells by many hits, and it is also assumed that a single ionization or ion cluster represents a hit. Many ionizations will occur within a target if hit by densely ionizing radiation (Fig. 3–5), and each of these ionizations will contribute an increment to the dose. If only one of these is required for inactivation, as assumed for viruses, the relative effectiveness per unit dose of these radiations will be low. Sparsely ionizing radiations will produce only a few ionizations within the target, however,

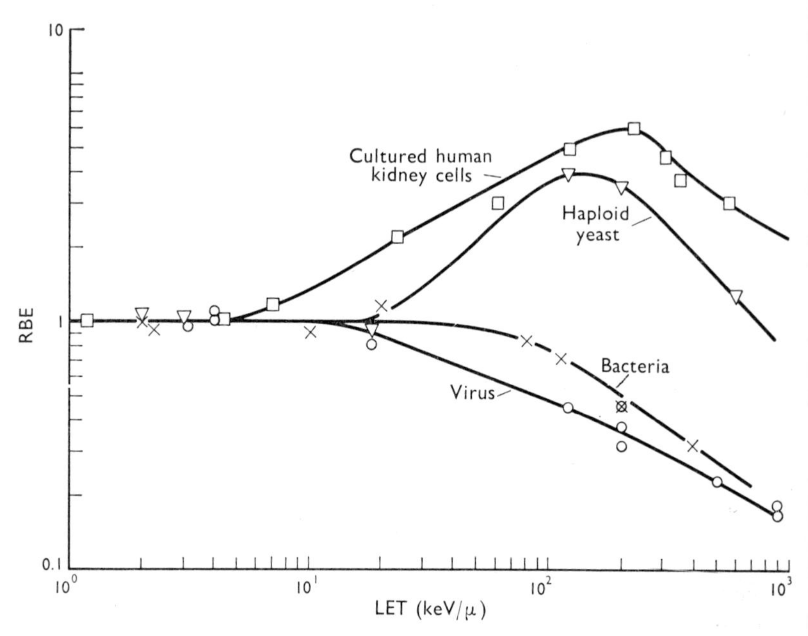

Fig. 3–6 The relationship between relative biological efficiency (RBE) and linear energy transfer (LET) for viruses, bacteria, and higher organisms. (Courtesy T. Brustad, *Radiation Research*, **15**, 139–158; T. R. Manney, T. Brustad, and C. A. Tobias, *Radiation Research*, **18**, 374–388; P. Todd, *Radiation Research, Suppl.*, **7**, 196–207, and Academic Press)

because the distances between consecutive ion clusters are great compared with the dimensions of the target (Fig. 3–5). Such radiations will therefore be efficient, and RBE will decrease with increasing LET. By the same argument, if many ionizations are needed to inactivate the target, as is assumed to be the case in mammalian cells, the RBE will increase with increasing LET, at least to the point where the optimum number of ionizations are produced within the target. Beyond this, RBE will be expected to decrease.

A multitarget hypothesis proposed by NEARY (1965) suggests that the cells of higher organisms are killed by damage within two targets, and according to this idea densely ionizing radiations are efficient because there is a high probability that the track of a single particle will simultaneously damage both of them. This is much less likely with sparsely ionizing radiations which kill the cell only if separate tracks pass through each target, a relatively infrequent occurrence. Finally it has also been proposed that higher organisms possess a greater capacity to repair damage than bacteria or viruses, but that the damage produced by high LET radiations is much less susceptible to repair. It is unlikely that any one of these explanations is adequate by itself, but some combination of all three is likely to apply in different circumstances.

Information of three kinds must be obtained before more detailed analyses of survival curves and their LET dependence can be attempted. First, more must be found out about the distribution of ionizations within irradiated materials. Average LET values are an inadequate description of radiation quality because all radiations have a range of LET values, the incident particles and ejected electrons becoming more densely ionizing as they slow down within irradiated materials. Second, the physical nature of a 'hit' must be determined, possibly by chemical analysis of irradiated DNA and correlated biological studies. Last, the ways cells repair their damaged DNA must be better understood.

The Genetic Effects of Ionizing Radiations 4

4.1 Chromosome aberrations

In addition to the efficiency with which they kill cells and prevent unlimited cell division, ionizing radiations are also very effective at producing genetic damage in living things, that is at damaging the structure of chromosomes and producing heritable changes or gene mutations. The ability of ionizing radiations to produce gene mutations was first demonstrated by H. J. Muller in 1927 and a detailed study of chromosome damage was started at about the same time. Since most mutations have deleterious effects, this work was important because it emphasized that radiations present a hazard not only to the people exposed to them but also to the descendants of these people. Examination of the genetic effects of radiations has also been important for the insights it provides into the activity and organization of genes and chromosomes, while the ability to induce mutations has had a profound effect on the development of genetics.

Abnormal or *aberrant* chromosomes are usually detected by examining stained cells with the aid of a microscope and are therefore observed only in animals or plants which have large, easily distinguished chromosomes. These chromosomes are composed of nucleoprotein (see § 5.3) and it is not clear whether similar aberrations occur in organisms such as bacteria whose chromosomes consist only of DNA. Chromosome aberrations of all kinds occur spontaneously in a very small proportion of cells and irradiation merely increases this proportion rather than producing something entirely new. When induced in this way, they are one of the first signs of permanent damage and can be seen in cells going through their first mitotic (or meiotic) division after irradiation. At other stages in the cell cycle, chromosomes are too diffuse and uncoiled to be distinguished clearly.

Many kinds of aberration have been found, some of which are shown in Fig. 4–1, but all are of two main types; those which involve both sister chromatids at any one position along the chromosome, called *chromosome* aberrations, and those which involve only one of the sister chromatids, called *chromatid* aberrations. Chromosome aberrations are found in cells that were irradiated before the chromosomes had replicated themselves, which takes place in the middle of the interphase between successive cell divisions, while chromatid aberrations are found in cells irradiated after replication. Some cells irradiated after replication appear to contain chromosome type aberrations but these are due to simultaneous damage to both sister chromatids, and are called *isochromatid* aberrations. Both chromosome and chromatid aberrations can take the form of either *exchanges*, in which a

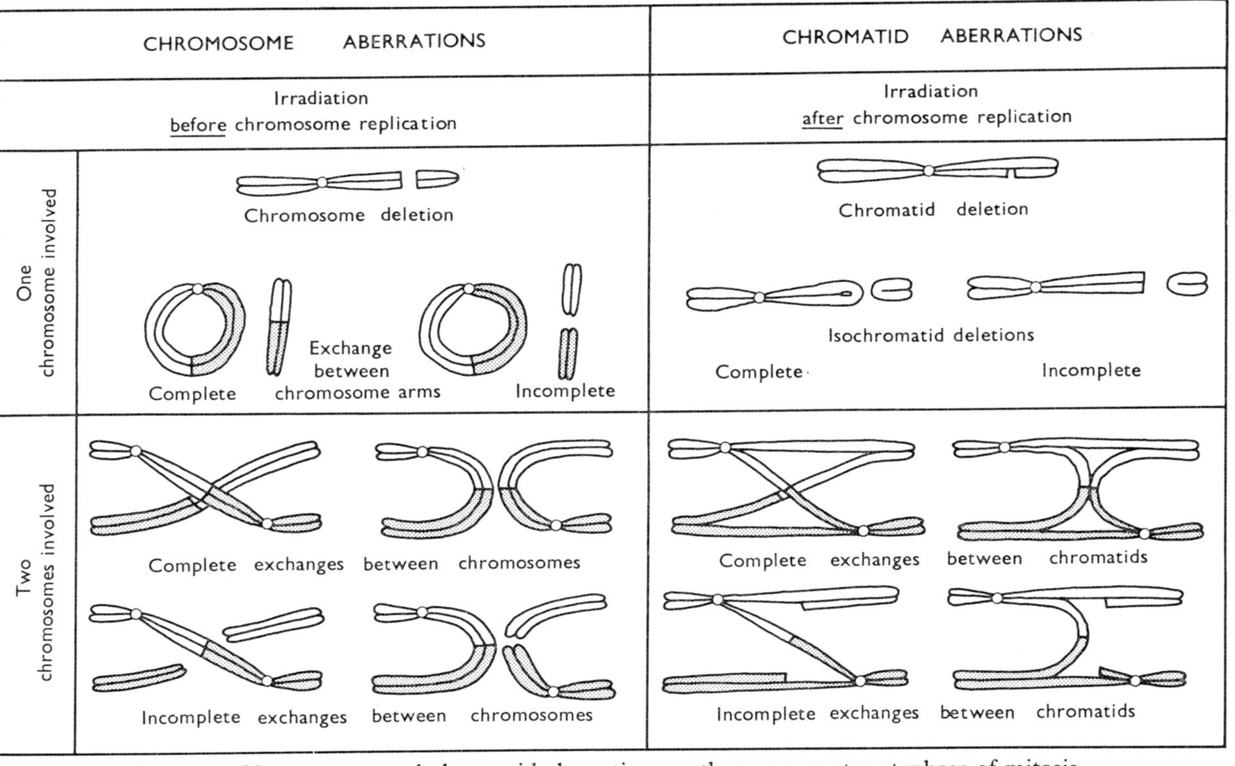

Fig. 4–1 Chromosome and chromatid aberrations as they appear at metaphase of mitosis.

portion of one chromosome or chromatid is wrongly attached to another, or of *deletions*, in which the chromosomes or chromatids are broken into two pieces (Fig. 4–1). Exchanges can be classified further according to whether they are complete or incomplete, and which way round the exchange takes place.

Two hypotheses have been put forward to explain the origin of aberrations. The 'breakage first' hypothesis proposed by SAX (1940) suggests, as its name implies, that radiation first breaks the chromosome thread, probably by the passage through it of a single ionizing particle. The two broken ends produced in this way then have three alternative fates. First, they may rejoin with one another to restore or restitute the normal chromosome. This is assumed to take place in most cases and such an event cannot, of course, be detected. Second, the two broken ends may persist in that state to produce a deletion. Third, if ends from other broken chromosomes or chromatids exist sufficiently close by, two or all of the four ends may rejoin wrongly to produce an exchange aberration. The essential features of this hypothesis are therefore that breakage and rejoining are independent events, separated in time, and that deletions require only one break produced by one 'hit' though the simplest exchange aberrations require two.

The 'exchange' hypothesis proposed by REVELL (1955) differs from this in two main respects. It suggests that *all* aberrations, including deletions, arise from an exchange process which is perhaps similar to meiotic crossing-over, and that the primary effect of radiation is not so much to break the chromatid as to initiate an exchange; that is breakage and rejoining are unified into one event. Chromatids in irradiated cells are thought to be damaged, though not broken, at many sites along their length and an exchange can be initiated if two of the sites are close enough together. When the two sites of damage are located on different, non-sister, chromatids this will eventually produce what is classified as an exchange aberration, but when they are on sister chromatids, or the same one, at a point where the chromatids loop across themselves, exchange can eventually produce what is classified as a deletion (Fig. 4–2). In some cases a tiny fragment is also expected but this may be too small to be seen. Since chromosomes are highly coiled structures they will loop across themselves at many points.

The two hypotheses differ most, therefore, in the way deletions are thought to occur. According to the 'breakage first' idea deletions are breaks which have failed to restitute, and arise as the consequence of a single hit, but according to the 'exchange' hypothesis they arise from exchanges within a chromosome loop and need two hits. It must be stressed that there is always a period between irradiation and division during which chromosomes cannot be examined, so that the correctness of one or other of these ideas cannot be established by direct observation, and the appearance of the aberrations offers little clue as to which is right. Even if cells are irradiated during division, aberrations can only be seen at the

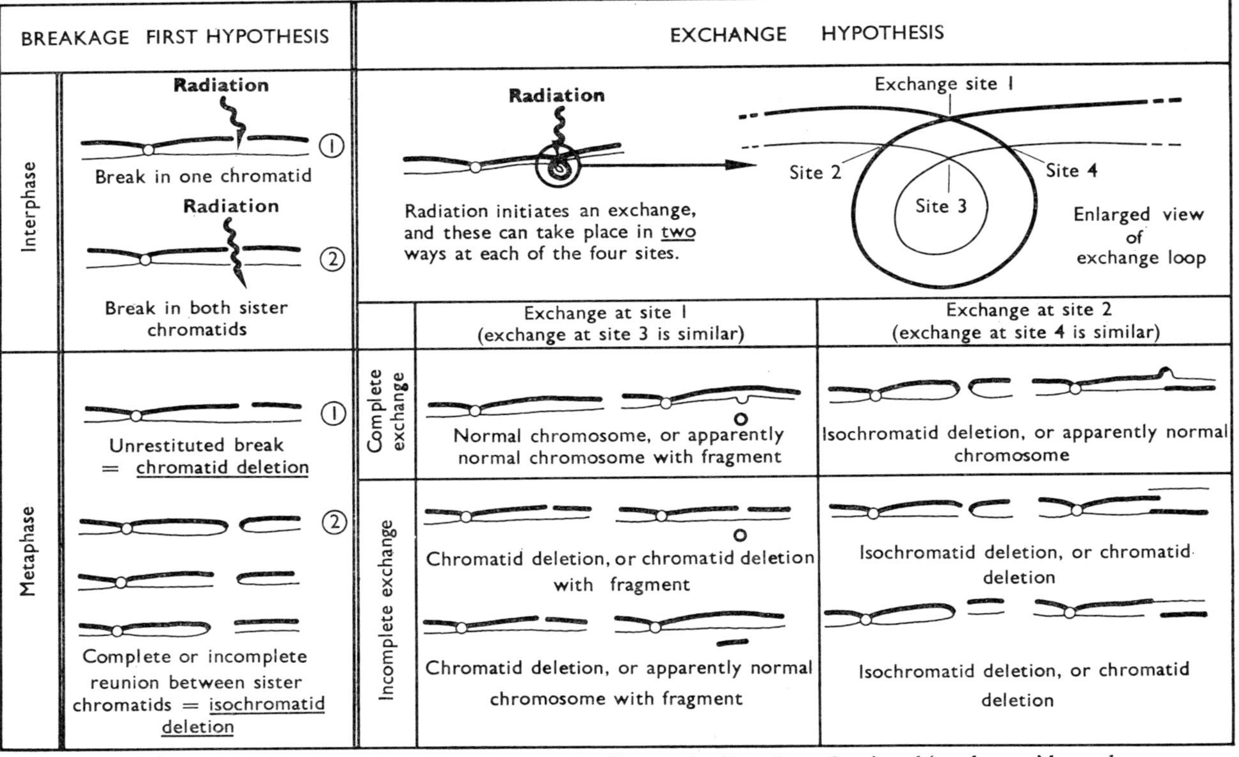

Fig. 4–2 The origin of chromatid deletions according to the 'breakage first' and 'exchange' hypotheses.

following division, either because they are concealed or because their formation is delayed.

Although posed as alternatives, it now appears likely that *both* hypotheses are correct (HEDDLE and BODYCOTE, 1970), and deletions probably arise sometimes in one way and sometimes in the other, the relative frequencies no doubt depending on the type of cell, the stage in the cell cycle irradiated, and so on. The first attempts to decide between the two ideas made use of dose–response relations which were interpreted on the basis of 'Hit' theory (see § 3.2). If a particular type of aberration always arises as the consequence of a single hit, or from damage produced by the track of a single ionizing particle, the average number of these aberrations per cell must show a simple linear relation with size of dose because the number of ionizing particles is simply proportional to dose. The relation will therefore be

$$\text{aberration yield} = kD^1 + C \quad \text{(one-track curve)}$$

where k is a constant, D the dose, and C a constant equal to the aberration frequency in unirradiated cells. The difference between this one-hit expression and that for loss of proliferative capacity (see § 3.2), where surviving fraction is *exponentially* related to dose, is merely a consequence of the different ways these injuries are observed. In the latter case it is only possible to observe the proportion of dead cells, that is those which contain hit targets, and no distinction can be made between cells hit once, twice, or more times. Cells containing different numbers of aberrations can be readily identified, however, and it is therefore possible to record the average number of aberrations per cell rather than only the number of cells containing aberrations. Nevertheless, aberrations, like lethal hits, are distributed between cells according to Poisson's formula and the proportion of cells free of aberrations is, like surviving fraction, exponentially related to dose.

If another type of aberration requires two hits, or more accurately hits from the tracks of two ionizing particles, their average number per cell will be proportional to the *square* of the dose; that is doubling the dose will quadruple their frequency. This is because radiations interact randomly with matter and the two hits are independent events. If x per cent of the cells receive a hit at the first site, and hits at the second site are equally frequent, x per cent of this x per cent, or x per cent squared of the cells will receive hits at both sites. Since the frequency of hits at any one site is simply a measure of the dose, the number of aberrations will be proportional to the dose squared.

$$\text{aberration yield} = kD^2 + C \quad \text{(two-track curve)}$$

Similar reasoning shows that if an aberration requires hits from the tracks of three ionizing particles, their frequency per cell will be related to the cube of the dose

$$\text{aberration yield} = kD^3 + C \quad \text{(three-track curve)}$$

The value of the constant k will vary according to the particular aberration concerned, the experimental conditions, and so on. C, the spontaneous aberration frequency, is usually so small that it can be ignored.

Simple 'hit' theory therefore suggests that the dose exponent, or the power to which the dose must be raised, in the curve which best fits the dose–response results gives a direct indication of the number of ionizing tracks whose damage produces the aberration in question. Although this is not always the same as the number of hits needed to produce the aberration, because a single densely ionizing track may produce more than one hit, the two tend to equal one another when sparsely ionizing radiations are employed. This happens because the track of a sparsely ionizing particle can rarely achieve more than one hit owing to the large distance between consecutive ion clusters. The shapes of the dose–response curves with sparsely ionizing radiations therefore provide an insight into the origin of the various types of aberration.

A survey of the first experiments of this kind (LEA 1946) showed that the results often agreed reasonably well with these simple expectations and this is also true, at least for certain kinds of aberrations, of more recent results such as those shown in Fig. 4–3. These data show that the yield of

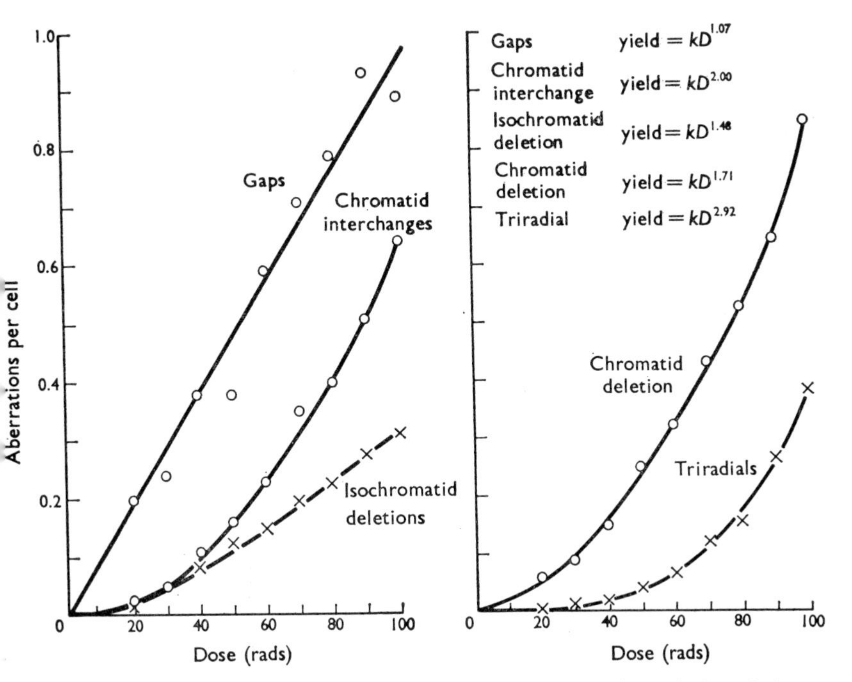

Fig. 4–3 Dose–response curves for chromatid aberrations induced by X-rays in the root-meristem of *Vicia faba*. (Courtesy S. H. Revell, *Mutation Research*, 3, 34–53, and Elsevier Publishing Co.)

simple chromatid exchanges is proportional to the square of the dose, hence suggesting they are 'two-hit' aberrations, while the yield of the more complex exchanges called triradials is nearly proportional to the cube of the dose, which suggests they require three 'hits'. For the purpose of deciding between the 'breakage-first' and 'exchange' hypotheses it is, however, the dose–response curve for simple deletions that is crucial. The results cited by LEA (1946) seemed to indicate that the yield of these aberrations was always proportional to the first power of the dose and if true this would support the 'breakage first' hypothesis. Simple chromosome deletions are very rare, however, and accurate estimates of their frequency are not usually obtained, while chromatid deletions are the most difficult aberration to recognize. After careful examination of irradiated cells in the root-tip meristem of *Vicia faba*, REVELL (1955) concluded that aberrations previously recorded as chromatid deletions were in fact of two kinds. More than 90 per cent of them were not real deletions at all, but non-staining gaps in the chromatids, which remained unbroken. The frequency of these was linearly related to dose (Fig. 4–3). The yield of true chromatid deletions by contrast increased in proportion to the 1.71 power of dose, an observation which can only be explained in terms of the 'exchange hypothesis'. Although the dose exponent was only 1.71, these aberrations are presumably 'two-hit' with some produced by a single track, perhaps the densely ionizing δ-rays (see § 2.2) or ends of the electron track which occur even with sparsely ionizing radiations. The same explanation probably applies to the isochromatid aberration results. A non-linear dose–response curve for simple chromosome deletions has also been found (BREWEN and BROCK, 1968). Others (e.g. CONGER, 1967) find few gaps and a linear dose–response for true chromatid deletions, and this is understandable if deletions are formed sometimes by breakage and at other times by exchange.

When densely ionizing radiations are used all dose–response curves become more linear, no doubt because a single densely ionizing track can produce more than one hit. At the same time such radiations are more efficient than the low LET variety, and RBE/LET relation for chromatid exchanges (SKARSGARD *et al.*, 1967) is remarkably similar to that for loss of proliferative capacity (Fig. 3–6), so that similar explanations may apply.

4.2 Gene mutations

Gene mutations can be recognized as changes in some aspect of the cell's structure or activities which is also expressed in the daughter cells of subsequent generations; that is the change is inherited and is a characteristic of the whole cell lineage. They are due to permanent changes in the genetic material and although this is usually replicated and transmitted from parent to daughters with great precision, very occasionally an error occurs, even without irradiation, and a mutant cell is produced. The mutations induced by radiation are no different from these spontaneous ones, but their frequency is much higher, by a thousandfold or more depending on dose. Without radiation, mutational changes in any given gene usually occur in less than one in a million cells.

Mutations in general are of many different kinds, and can be classified either according to types of alterations in the genetic material responsible for them or according to the kinds of change they produce in the cell, their phenotype. Some mutations are due to inclusion of an extra chromosome, or even whole sets of chromosomes, in the nucleus, while others arise from the addition or loss of a small part of a chromosome, that is, they are due to chromosome aberrations. These *chromosome mutations* involve changes in the number of genes in the nucleus. Yet other mutations are due to extremely small changes in DNA molecules such as the addition or loss of a single pair of nucleotide bases or the substitution of one kind of base pair for another. These are called *point mutations*, and involve changes in the nature but not number of genes.

It is rarely possible, however, to determine what kind of alteration has taken place in the genetic material of a mutant cell, so mutations are usually classified according to their phenotype and the way in which they are detected. *Biochemical* mutations are often studied in micro-organisms, and are recognized by the inability of the mutant to grow unless some compound, such as an amino acid or vitamin, for which the normal cell is self-sufficient, is added to the nutritive medium on which they are cultured. An enzyme concerned in the biosynthesis of the compound is usually missing or defective in these mutants, which therefore contain alterations affecting the gene responsible for the enzyme. *Reverse* mutation, back to the normal state of self-sufficiency, is very frequently studied with biochemical mutants. *Resistance* mutations are examined both in micro-organisms and cultured mammalian cells, and these allow the mutant to grow in the presence of a particular antibiotic, poison, or similar substance at concentrations which normally prevent growth or are lethal. *Lethal* mutations are sometimes studied in micro-organisms, but more commonly in higher animals, where they are induced in germ cells. *Visible* mutations, which change the morphology of the organism, are also studied mostly in higher animals or plants.

Micro-organisms or cultured cells are useful for mutation studies because large numbers of uniform cells, each a potential mutable unit, can be produced and these are necessary to estimate the frequency of mutants which, even after irradiation, is usually very low. A good example of their use, and also of the use of 'hit' theory in mutation research, is provided by the work of DE SERRES and his colleagues (1967), who have studied mutations affecting two adjacent genes in *Neurospora crassa* which impair the ability of this fungus to synthesize adenine. When sparsely ionizing X-rays were used, the relation between mutation frequency and dose was best fitted by a curve which had a dose exponent between one and two. As in the case of chromosome aberrations, this implies that some of the mutations arise from damage produced by one track and some from damage produced by two tracks. The relative contributions of these events to the yield of mutations can best be estimated by fitting a quadratic expression to the results, with terms depending on dose and dose squared, rather than the dose exponent type of expression and this was found to be

$$\text{yield (mutants per } 10^6 \text{ surviving cells)} = 0.3 + 5.39x + 0.16x^2$$

where x is the dose in kilorads. This shows that in every million surviving cells there are, on average, 0.3 spontaneous mutants, 5.39 one-track mutants per kilorad, and 0.16 two-track mutants per kilorad squared.

Superficially, all of the induced mutants looked very similar to one another but it was of interest to discover if the one- and two-track mutants arose from different kinds of changes in the DNA. By using an ingenious system of test crosses it was in fact possible to show that the one-track types were point mutations and the two-track types chromosome mutations, in which a piece of chromosome containing the adenine loci had been deleted and lost from the nucleus. It is likely, therefore, that point mutations are one-hit, and chromosome mutations two-hit, events. When densely ionizing radiations were used instead of X-rays (DE SERRES *et al.*, 1967), the two-track component became smaller, because a single track could produce both hits required for the deletion-type mutations. At the same time, high LET radiations were more efficient at producing mutations than X-rays, and the change in RBE with increasing LET (see § 3.3) was nearly identical to that for loss of proliferative capacity. The same result was also found by MORTIMER *et al.* (1965) for point mutations in yeast.

The results from many experiments support the conclusion that the frequency of point mutations is directly proportional to dose, at least within the range of doses at which this frequency can be estimated. It is not known whether it is also true at very low doses, where the induced mutation frequency is too low to be estimated. The question is important, however, because such doses may be significant when attempts are made to assess radiation hazards to human populations. The relative proportions of chromosome and point effects varies not only with size of dose, but also depends on the kind of cell or organisms used, the type of mutations studied, the conditions of growth, and so on. In some cases, deletion-type mutations are recessive lethals and so could never be recovered from haploid cells. In higher plants, however, most mutations are thought to be due to small deletions. Point mutations are more common in animals and micro-organisms. For these and other reasons, the shape of dose–mutation frequency curves can vary widely from experiment to experiment.

4.3 The relation between genetic and lethal damage

As discussed in Section 3.2, there is now a considerable amount of evidence which suggests that cells lose the ability to proliferate because of damage to their DNA. Since genetic damage presumably also arises from alterations within DNA molecules, the question is raised of whether the same kinds of alteration are involved in each case and whether the difference between the lethal and genetic effects of radiations resides merely in the way that they are observed. In general, it might be expected that lethal damage depends on a great variety of alterations in the DNA located at any one of a large number of sites, while genetic damage probably depends on a smaller range of alterations, located in the case of mutations at only a few sites. Nevertheless, most chromosome and chromatid aberrations interfere with the exact and orderly transmission of the genetic material to daughter nuclei and could therefore represent at least one of the causes for loss of proliferative capacity. Lethal point mutations on the other hand are too rare to account for more than an insignificantly small proportion of dead cells. Many aberrations lead to the loss of genetic material during division because, as can be seen in Fig. 4–1, they involve the formation of acentric fragments, that is pieces of chromosome lacking a centromere. Normal chromosomes each possess one centromere somewhere along their length and this structure is concerned in their anaphase movement. Acentric fragments cannot move normally and therefore usually fail to be included in either of the daughter nuclei. Although most cells can survive the loss of a small amount of genetic material, the loss of a large amount is almost bound to be lethal. Similarly, dicentric aberrations, which have two centromeres on the same chromosome, impose difficulties during division.

Because of this, the suggestion was once made that loss of proliferative capacity was *always* due to the formation of aberrations, but this no longer seems likely (DAVIES and EVANS, 1966). Although aberrations may be responsible for the death of cells which attempt to divide at least once after irradiation, they can hardly account for the death of those which do not divide. Moreover, those which do not normally divide, such as the differentiated cells which make up the bulk of most multicellular organisms, should always be much more radioresistant than their dividing counterparts, but although this is often the case it is not always so. It must be concluded, therefore, that aberrations can at best account for loss of proliferative capacity in only some of the cells which fail to survive the irradiation.

The Biochemical Consequences of Irradiation — 5

5.1 Radiation chemistry

Since it is likely that damage to DNA is responsible for loss of proliferative capacity in irradiated cells, as well as for gene mutation and chromosome aberrations, it is important to examine the chemical alterations that radiation can produce in these macromolecules and the reactions which give rise to them. This is not easy to do directly, however, because most analytical techniques are too insensitive to detect small, though biologically significant, modifications in such large molecules, and when DNA is irradiated within cells the reactions which produce these changes occur in chemical circumstances that are both complex and unknown. Instead, the problem must be approached in several complementary ways, including experiments with simpler organic and inorganic compounds as well as with purified DNA in the test-tube. Finally, it is also necessary to try to determine which of the many radiation-induced alterations in DNA are responsible for cell death, chromosome aberrations, and so on.

Chemical reactions occur within irradiated materials because the radiation provides the *activation energy*. Chemical reactions in general may either release energy as in combustion, or require a net addition of energy as in the synthesis of organic compounds. In each case, however, there is an energy barrier which must be overcome before the reaction can proceed and relatively large amounts of energy, the activation energy, must be available temporarily to overcome the barrier. This can be supplied either in the form of heat, the rapid oscillation of all of the reactant molecules, or as localized units of energy, activating only some of the many molecules which could react, so that the reaction takes place in 'cold' surroundings. Radiation supplies activation energy in this second way, and the units reside in ionized and excited molecules. The molecules which are eventually altered are not necessarily those originally ionized, however, because energy can be passed from one to another. Although radiation, particularly of the densely ionizing varieties, causes local heating, the heat is dissipated too quickly to allow reactions to occur and once spread throughout the material the heating effect of radiation is very small, about two-thousandths of a degree centigrade per thousand rads.

Most chemical reactions require an activation energy of the order of one to ten electron volts, and this is easily supplied by one ionization which is roughly equivalent to thirty electron volts. The number of reactions per unit volume is proportional to the amount of energy deposited, that is to dose, but even though very effective only a small proportion, probably no more than a quarter, of the absorbed energy is finally trapped in the form of stable chemical change, the remainder being harmlessly dissipated as heat.

The sequence of events leading to such chemical alterations can be divided into two main stages. During the *physicochemical stage* the highly unstable ionized and excited molecules undergo reactions, either spontaneously or in collisions with other molecules, to give rise to inherently stable but very reactive entities called *free radicals*. This takes about 10^{-10} s, during which the thermal equilibrium of the irradiated material is restored. The ions produced by irradiation must not be confused with the stable ions in dissociated salts. They are more correctly called free radical ions because, like free radicals, they possess an unpaired electron in their outer electron shell. Free radical ions and electronically excited molecules may both give rise to free radicals, but they do so with very different probabilities. Excited molecules, especially excited macromolecules, have a variety of ways of losing the excess energy they contain without undergoing chemical alteration, but ionized macromolecules almost invariably become involved in some kind of reaction. Consequently, even though excited molecules may be several times more abundant than ionized ones, they probably have much less radiobiological significance. The different ways in which free radicals can be formed from free radical ions is not well understood, particularly for the complex macromolecular ions, but some are unstable and dissociate into two pieces, one of which may be a free radical and the other an ion. Dissociation is particularly likely if a positive ion captures an electron or if positive and negative ions combine, and in these cases two free radicals are formed.

During the following *chemical stage*, which usually lasts 10^{-6} s, free radicals react with one another or with normal molecules. Although inherently stable, free radicals are usually very reactive because of their unpaired electron, and quickly become involved in reactions. The speed with which they do so depends largely on their mobility, however, and the mobility of the molecules around them. Most cells contain water through which small free radicals readily diffuse, and being so reactive they rarely diffuse further than a few nanometres from the original site of ionization. Certain dormant cells in seeds or spores contain little water, however, and some free radicals may persist for months in these, although they quickly disappear when water is imbibed.

5.2 Direct and indirect action

Since water accounts for 75 per cent or more of the weight of most cells, roughly three-quarters of the ionizations produced in them occur in water molecules and the highly mobile free radicals that are created may eventually attack vital macromolecules, such as DNA. In principle, therefore, radiation can damage macromolecules in two ways; by *direct action*, in which the macromolecules themselves are ionized, and by *indirect action*, in which water or solute molecules are ionized and then produce free radicals which damage macromolecules. In practice, it is usually difficult to discover the relative proportions of direct and indirect damage because even if cells contain a lot of water, the particular macromolecules concerned may be isolated from this water, or dissolved substances may react preferentially with the mobile free radicals and so protect the macromolecules from damage.

The main products of the radiolysis, or radiation-induced breakdown, of water are hydrogen and hydroxyl free radicals,

$$H_2O \longrightarrow H\cdot + OH\cdot$$

The dot, as in $OH\cdot$, indicates the unpaired electron in these radicals which can be formed directly by dissociation of excited water molecules or by their ionization.

$$H_2O \xrightarrow{\text{ionization}} H_2O^+ + e^-$$

When the ejected electron (e^-) has slowed down sufficiently, it can be captured by a water molecule to form a negatively charged ion.

$$H_2O + e^- \longrightarrow H_2O^-$$

The positive and negative free radical ions are both unstable, and each dissociates to form a stable ion and a free radical,

$$H_2O^- \longrightarrow OH^- + H\cdot$$
$$H_2O^+ \longrightarrow H^+ + OH\cdot$$

The stable hydrogen (H^+) and hydroxyl (OH^-) ions combine to form water, so that overall the reaction gives hydrogen and hydroxyl free radicals. Electron capture by neutral water molecules is, however, a relatively slow process and the electron may become hydrated, that is loosely associated with surrounding water molecules and exist as a separate entity for a relatively long time ($\leqslant 10^{-6}$ s). The hydrated electron (e^-aq) reacts very quickly with hydrogen ions to give hydrogen atoms (free radicals),

$$H^+ + e^-aq \longrightarrow H\cdot$$

but at the pH normally encountered within cells may persist long enough to attack organic substances. $H\cdot$ and e^-aq are both powerful reducing agents and the kinds of damage they produce are very similar.

$H\cdot$, $OH\cdot$ and e^-aq are all very reactive, and during the course of diffusion distribute the absorbed energy to solute molecules, both organic and inorganic, with high efficiency. During this process the primary free radicals may give rise to secondary, somewhat less reactive, radicals and molecules which in turn are capable of attacking macromolecules. For example, $H\cdot$

and e^-aq both react with oxygen to form the hydroperoxy radical $HO_2\cdot$:

$$O_2 + H\cdot \longrightarrow HO_2\cdot$$
$$O_2 + e^-aq \longrightarrow O_2^-$$
$$O_2^- + H^+ \longrightarrow HO_2\cdot$$

The primary free radicals also react with each other:

$$H\cdot + OH\cdot \longrightarrow H_2O$$
$$H\cdot + H\cdot \longrightarrow H_2$$
$$OH\cdot + OH\cdot \longrightarrow H_2O_2$$

Hydrogen peroxide, which can also form by the combination of two hydroperoxy radicals,

$$HO_2\cdot + HO_2\cdot \longrightarrow H_2O_2 + O_2$$

is a stable molecule but can oxidize and therefore damage organic substances. The probability that free radicals react with one another, rather than with solute molecules, depends largely on the spacing of the ionizations in the water. With sparsely ionizing radiations the ionizations, and therefore the free radicals, are generally formed so far apart that they have little chance of meeting one another before encountering solute molecules. With densely ionizing radiations like α-rays, however, the local concentration of radicals is high. As a consequence, relatively high concentrations of hydrogen peroxide are formed around the tracks of α particles and may contribute to the damage which this radiation causes.

The attack on proteins, nucleic acids, etc., by free radicals leads to the abstraction or addition of hydrogen to these molecules and the formation of a macromolecular radical. If RH represents the protein or nucleic acid, the initial reactions are:

$$RH + H\cdot \longrightarrow R\cdot + H_2$$
$$RH + OH\cdot \longrightarrow R\cdot + H_2O$$
$$RH + H\cdot \longrightarrow RHH\cdot$$
$$RH + OH\cdot \longrightarrow RHOH\cdot$$

These secondary radicals ($R\cdot$ etc.) are probably quite similar to those formed by the direct ionization of the macromolecule:

$$RH \xrightarrow{\text{ionization}} RH^+ + e^-$$

$$RH^+ \xrightarrow{\text{dissociation}} R\cdot + H^+$$

The site within the molecule at which radical formation takes place is not necessarily that from which the electron was lost. Immediately after ionization there is a very rapid redistribution of electrons within the macromolecule and the electron deficiency is localized in the region best able to support it. The electron deficiency may even be 'exported' to an adjacent molecule which therefore protects its neighbour from damage.

Finally, the macromolecular radicals, resulting either from the direct or from the indirect action of radiation, react with various solutes or with each other in such a way that the biological function of the molecules is often destroyed. Typical reactions are:

$$\begin{array}{rcll} R\cdot & \longrightarrow & A+B\cdot & \text{direct dissociation} \\ R\cdot+O_2 & \longrightarrow & ROO\cdot & \multirow{2}{*}{\text{oxidation}} \\ RHH\cdot+O_2 & \longrightarrow & RHHO_2\cdot & \\ R\cdot+R\cdot & \longrightarrow & RR & \text{crosslinking} \\ R\cdot+H\cdot & \longrightarrow & RH & \multirow{2}{*}{\text{restitution}} \\ R\cdot+XH & \longrightarrow & RH+X\cdot & \end{array}$$

In these reactions the active molecule is again represented by RH. In the cases of dissociation, oxidation, and crosslinking the biological function of the molecule is often impaired or lost. The peroxy radicals formed on oxidation are usually unstable and dissociate. The last two reactions lead to restitution of the original molecule which is therefore 'repaired'. XH in the last reaction can represent a variety of substances, the best known of which are those containing the sulphydryl group SH (see Chapter 6).

5·3 Radiation damage in DNA

The DNA in irradiated cells can take part, no doubt, in any one of the different reactions outlined above, but because of the difficulties of working with such large and complex molecules their relative importance, and the alterations to which the reactions give rise, are unknown. Since all are about two nanometres wide, the size of DNA molecules can be appreciated from their length. Bacteria contain a single molecule which is up to one millimetre in length, while some higher plants have tens or hundreds of molecules with a total length of ten metres. DNA in cells other than those of bacteria or blue–green algae is also tightly bound to protein to form a nucleo-protein complex. Each DNA molecule consists of two chains of nucleotides coiled helically around one another in the manner first proposed by Watson and Crick (Fig. 5–1). The linear integrity of the molecules depends on the two sugar–phosphate chains and these are joined by hydrogen bonds between complementary pairs of bases which project from the chains. Only adenine–thymine or guanine–cytosine pairing is allowed, and these four bases represent the 'letters' of the code used to store genetic information.

When purified DNA is heavily irradiated in the test-tube, chemical alteration in all parts of the molecule can be detected including deamination, dehydroxylation or loss of bases, oxidation of the sugar, loss of inorganic phosphate, and breakage of the bonds between sugar and phosphate (GINOZA, 1967). When very high doses are used, however, it is always possible that the results are not relevant to situations where the much lower doses typical of biological experiments are used. Only chemical

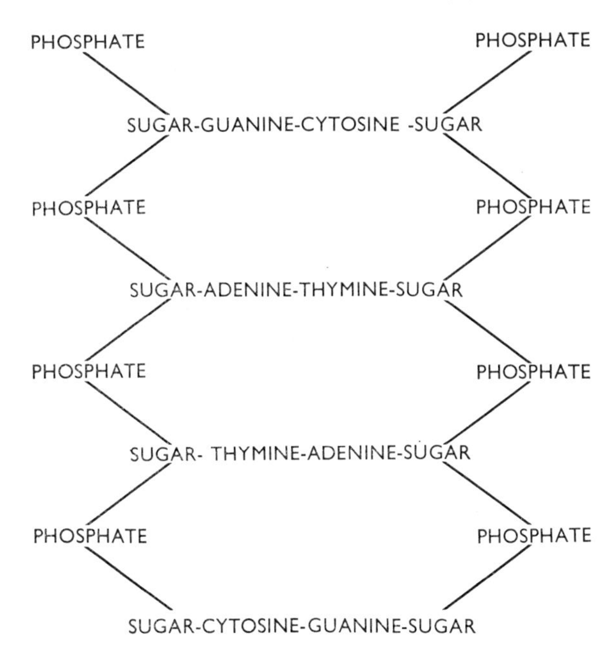

Fig. 5–1 The arrangement of subunits within DNA.

alterations which change the physical structure of DNA can be detected at lower doses, using the sensitive methods of the polymer chemist, and such physical alterations included cross-linking within or between molecules and breakage of the sugar–phosphate chain (Fig. 5–2). Because DNA has two chains, breakage in only one of them does not disrupt the molecule as a whole and if randomly distributed, many breaks can accumulate before the DNA is degraded into two or more pieces. When DNA is exposed to indirect action, the free radicals probably attack randomly and break only one chain at a time. Similarly it is likely that the direct action of sparsely ionizing radiations only breaks one chain at any given position, but densely ionizing radiations can break both chains at the same point. Degradation of the DNA into two or more pieces only occurs if both chains are broken at the same, or nearly the same point (Fig. 5–2). Since genetic information may be lost when DNA is degraded, double chain breaks are likely to be a particularly significant form of damage.

Because the molecules are so long they are very easily broken when DNA is chemically extracted from cells, particularly after irradiation. MCGRATH and WILLIAMS (1966), however, devised an extremely gentle method of extraction and using this were able to show the presence of breaks in DNA irradiated within the cell. Other attempts to assess the biological significance of the various kinds of damage to DNA have made

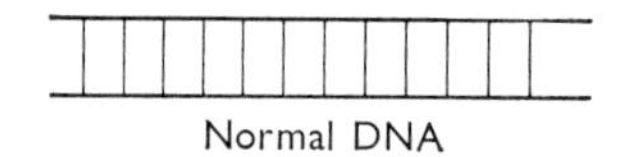

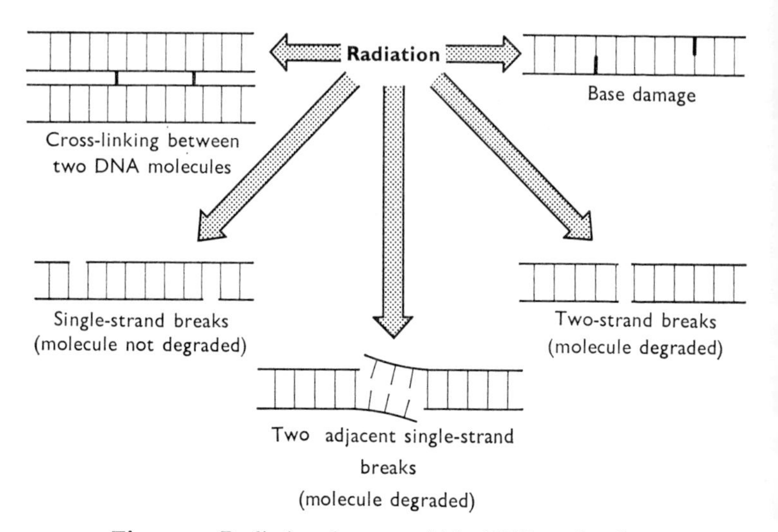

Fig. 5–2 Radiation damage within DNA molecules.

use of viruses because they are very much less complex than cellular organisms and consist essentially of relatively short pieces of nucleic acid wrapped in a protein coat. They possess no metabolic activity outside their host cells, and in this state can be irradiated under well defined conditions. Inactivation depends almost entirely on damage to DNA and the proportion of survivors can be estimated from the number of viruses which successfully multiply in their host cells.

Some viruses contain single-stranded DNA rather than the more usual double-stranded variety, and these can probably be inactivated by a single ionization or attack by a single free radical, anywhere within the DNA (GINOZA, 1967). If so, *any* chemical change in the DNA is lethal. This is not true for viruses which contain double stranded DNA, however. FREIFELDER (1968a), who studied four viruses of this sort, simultaneously measured the proportion of 'surviving' viruses and the proportion which contained DNA free of double strand breaks after irradiation with several low doses (Fig. 5–3). If these viruses were 'killed' only as a result of double strand breaks, the proportion of surviving particles should exactly equal the proportion of viruses containing unbroken DNA at any given dose; that is the dose–survival and dose–unbroken DNA graphs should exactly superimpose, but it can be seen that this is not the case. Although the two

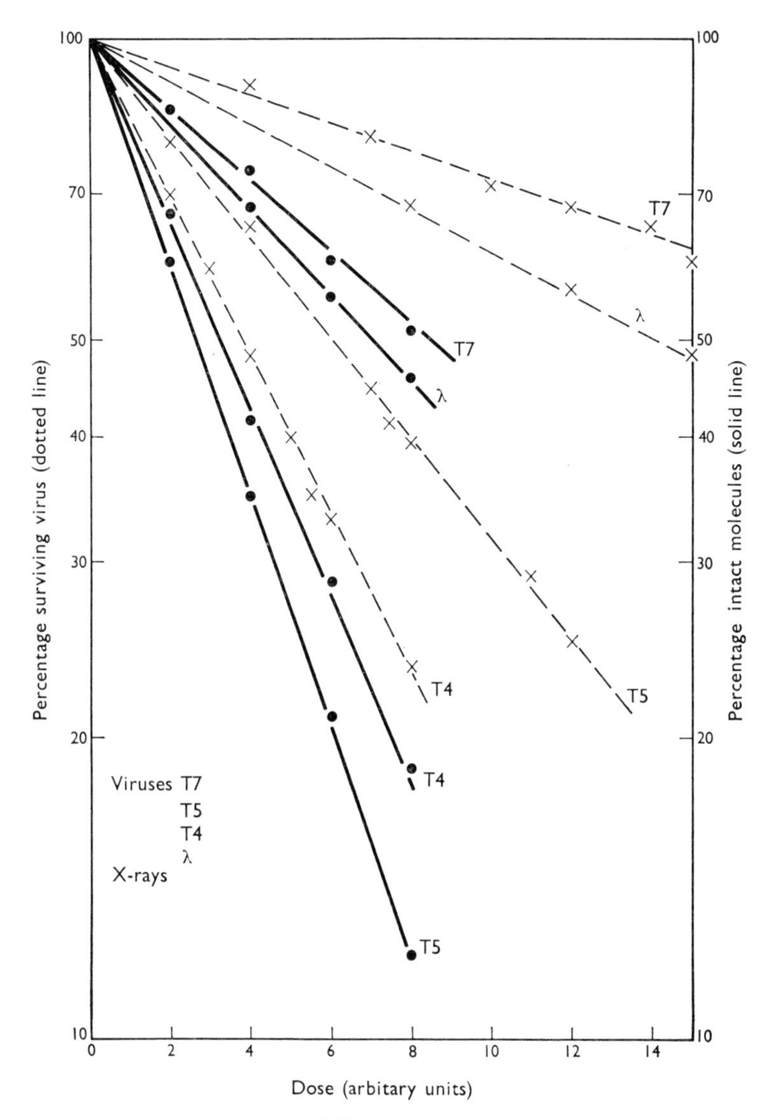

Fig. 5–3 Radiation-induced killing and damage to DNA in four viruses, T4, T5, T7, and λ. Dotted lines show proportion of viruses surviving various doses of X-rays and solid lines the proportion of DNA molecules without double strand breaks. Each virus has only one molecule of DNA. (Courtesy D. Freifelder, *Virology*, **36**, 613–619, and Academic Press)

lines for the T4 virus are quite similar, the dose required to inactivate a given proportion of the other three viruses is only about half the dose needed to break the DNA in the same proportion of particles, and therefore only half the dead viruses contain double strand breaks. The other half of the viruses may have been inactivated by single strand breaks, but as these are about ten times as frequent as double breaks, damage to the purine or pyrimidine bases, which cannot be estimated directly, is more likely. Single strand breaks also occur in a far higher proportion of cells than are killed in the bacterium *Escherichia coli* (FREIFELDER, 1968b), and it was concluded that these too were killed by a mixture of double strand breaks and damage to purine or pyrimidine bases.

Modification of Cellular Injury 6

6.1 Chemical and physiological modification

In Chapters 3 and 4, emphasis was laid on the physical stage (Fig. 1–1) in the development of radiation damage, either loss of proliferative capacity of genetic damage, and the nature of the events at this stage was analysed by varying the amount of energy absorbed, the radiation dose, and the distribution of the absorbed energy within irradiated cells, the radiation quality. The physical stage is of course only the first in the process, and the nature of the events in the chemical and physiological stages which follow can be analysed in an analogous way by varying the chemical composition of cells and their physiological or metabolic state. Cellular damage can be modified in a great variety of ways, including treating the cells with chemical substances or drugs and altering their supply of nutrients or their growth conditions, and experiments with these modifying treatments provide a powerful means of bridging the gap between observations on cells and those on macromolecules, discussed in Chapter 5.

In practice it is not always a straightforward matter to decide whether a treatment that modifies radiosensitivity does so by changing the chemical reactions which lead to macromolecular damage, or whether it changes the physiological responses of the cell to the presence of such damage. In general, however, action at the chemical and physiological stages can be distinguished in two ways. First, because chemical reactions induced by radiation take place very quickly, and are usually complete within a millionth of a second after energy absorption, treatments that act at the chemical stage are only effective if applied *during* the course of irradiation. Physiological responses to the presence of radiation damage are much slower and treatments which act at this stage are effective *after* the irradiation is over. One important qualification must be made, however. Free radicals may persist for long periods in 'dry' cells (see p. 35), up to the time water is imbibed, while such cells have little or no metabolic activity until water is present. Second, treatments which act at the chemical stage are usually *dose-modifying*. This means that the dose–response curves obtained with and without the treatment can be superimposed if the dose scale in one or other of the graphs is multiplied by a constant factor. Treatments which act at the physiological stage often fail to be dose-modifying.

6.2 Protective and sensitizing substances

Many substances are known to act at the chemical stage, some of which *protect* cells from radiation damage and others which *sensitize* them, or enhance the damage. Of these oxygen, water, and cysteamine will be chosen to illustrate their action and interrelationships.

Cells containing oxygen are more radiosensitive than those in which the oxygen has been replaced by an inert gas such as nitrogen or helium, and to achieve this sensitization the oxygen must be present within the cell during

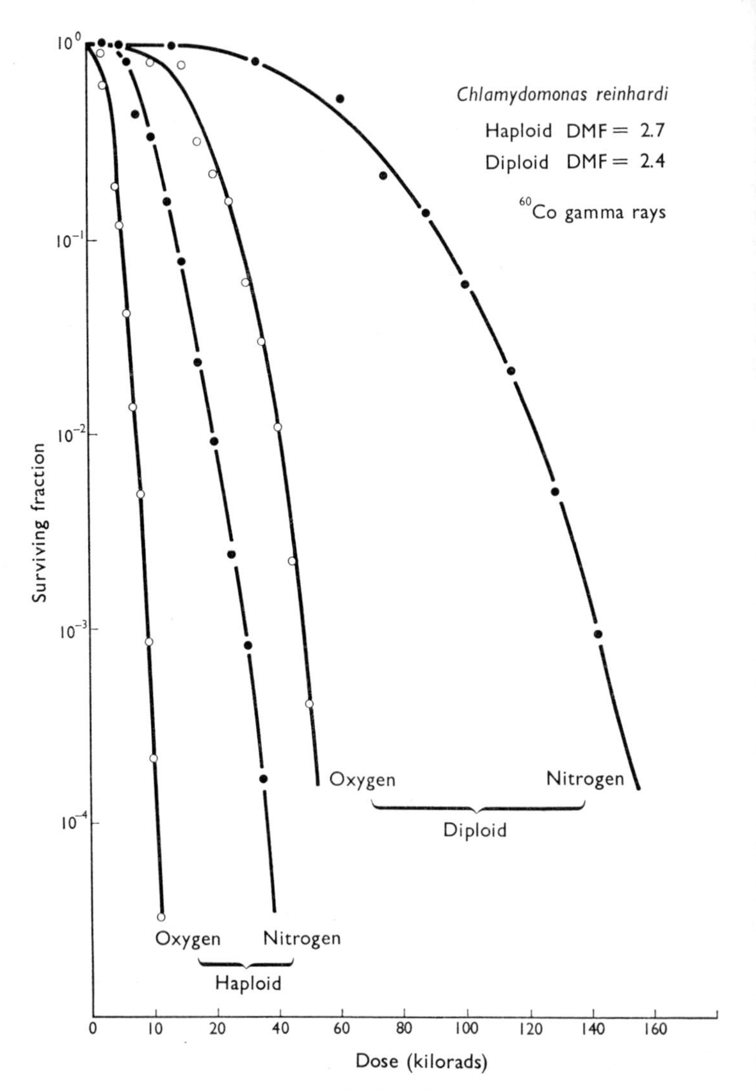

Fig. 6–1 Survival curves for haploid and diploid cells of the green alga *Chlamydomonas reinhardi*, irradiated in the presence of oxygen or nitrogen. (Courtesy D. R. Davies (unpublished data))

the actual irradiation period. Within the limits imposed by experimental technique, treatment of actively metabolizing cells with oxygen either before or after irradiation has no effect, and using a very ingenious method HOWARD-FLANDERS and MOORE (1958) have in fact shown that sensitization fails to occur even if oxygen enters cells no more than 0.02 s after the end of a very short irradiation treatment. As can be seen from Fig. 6–1, which shows the effect of oxygen on the survival of haploid and diploid cells of the green alga *Chlamydomonas reinhardi*, the presence of oxygen is *dose-modifying*, that is, there is a constant ratio between the doses required to reduce survival to any given level in the presence and absence of oxygen. This ratio, or dose-modifying factor (DMF), is about 2.7 in the haploid cells and 2.4 in the diploids, which means that radiation is 2.7 or 2.4 times less effective in killing these cells when oxygen is absent.

The actual size of the DMF for oxygen, which commonly lies between 2.5 and 3.0 but has been known to be as large as 5.0 or more, depends on many factors, such as the concentration of oxygen in the cell, the type of radiation used, the kind of damage studied, the type of cells irradiated and the conditions in which they are grown. Figure 6–2 shows how the DMF for loss of proliferative capacity depends on the concentration of dissolved

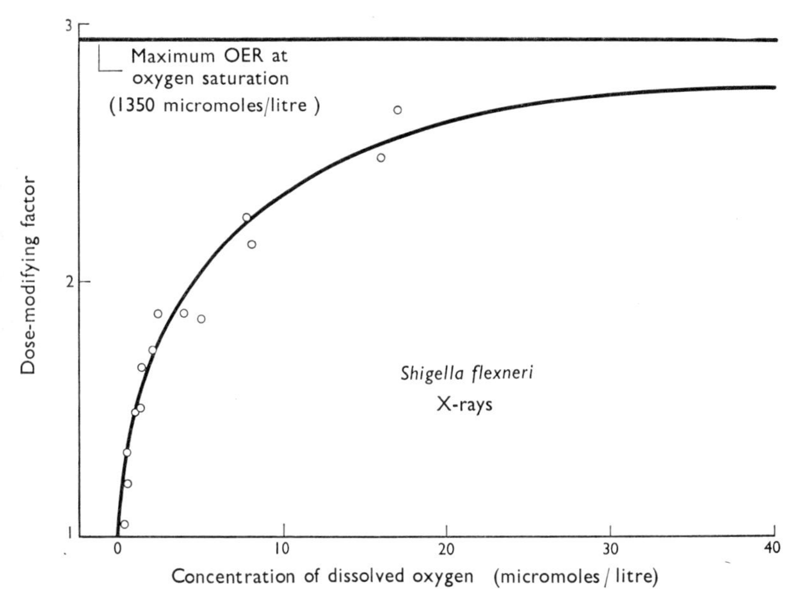

Fig. 6–2 The relationship between oxygen enhancement ratio (OER) and oxygen concentration in the bacterium *Shigella flexneri*. (Courtesy P. Howard-Flanders and T. Alper, *Radiation Research*, 7, 518–540, and Academic Press)

oxygen in the bacterium *Shigella flexneri*, and very similar curves have been found for other kinds of damage, such as production of chromosome aberrations. The maximum size of the oxygen DMF is greatest with sparsely ionizing radiations and decreases with increasing ion density, the oxygen effect often disappearing altogether at very high LET values (e.g. TODD, 1967). The maximum size of the oxygen DMF may also be rather smaller for the induction of some point mutations, than for survival or aberration production. In general the largest oxygen effects are found when cells are grown in conditions which make them most radioresistant, and the DMF becomes progressively smaller as cells become more sensitive.

A number of organic compounds which contain sulphur, particularly those which possess a sulphydryl group (SH), protect cells against radiation damage. Some of these are normal constituents of the cell, and sulphydryl-containing proteins, for example, are found in the chromosomes of many organisms. These naturally occurring protective substances are no doubt quite effective in protecting cells against some of the damaging effects of radiations, but it is possible to obtain additional protection by treating cells with other kinds of sulphydryl compound which they absorb. The most widely studied of these is *cysteamine*, which is the decarboxylated derivative of the amino acid cysteine, itself a radioprotector.

$$SH{-}CH_2{-}CH_2{-}NH_2 \qquad \text{Cysteamine}$$

Cysteamine protects against all kinds of damage in almost all types of organism, both in the presence and absence of water. Since it is effective only if present in the cell at the time of irradiation and can protect viruses, enzyme molecules, and even synthetic polymers against radiochemical alteration, it must act at the chemical stage rather than by altering cellular metabolism. Apart from the fact that it has the opposite effect, the action of cysteamine often parallels that of oxygen. The effect of cysteamine is usually dose-modifying, with DMF's which lie in the same range as those for oxygen, and the size of the DMF also depends in a similar way on intracellular concentration. The protective effect of cysteamine is greatest with sparsely ionizing radiations, becoming smaller at high LET values, and usually is also greater in the presence than in the absence of oxygen.

Since oxygen and cysteamine are effective only if present in the cell during the irradiation period, they must produce their effects by participating in the chemical reactions leading to macromolecular damage, but it is usually difficult to study such reactions because of the speed with which they take place. These difficulties can be overcome, at least in part, by using 'dry' cells which only contain up to 10 per cent by weight of water, rather than the more normal 70–80 per cent. Material of this kind occurs naturally, in the form of seeds and spores, and can be obtained by artificial drying. Some of the radicals induced by radiation persist for a long time in cells and since they are metabolically inactive, 'dry' cells can be exposed to toxic substances and conditions, such as high or low temperatures, which would kill normal materials. Radiation-induced chemical reactions can therefore be studied by the methods of physical chemistry while the consequences of the reactions can be assessed biologically after the cells have absorbed water and resumed active metabolism.

These properties of 'dry' cells have been exploited by Powers and his colleagues (POWERS, 1965) in a series of experiments, using dried spores of the bacterium *Bacillus megaterium* and dried cells of the bacterium *Staphylococcus aureus*, in which they examined the effects of oxygen, water, and hydrogen sulphide (H_2S) on loss of proliferative capacity. H_2S, the simplest of the sulphydryl compounds, is far too toxic to be tested on normal cells, but is a radioprotector in 'dry' cells, from which it can be removed before metabolism is resumed.

As found previously with seeds, oxygen can sensitize these 'dry' cells not only if present during the actual period of irradiation but also in the period following irradiation up to and including the time of hydration. On the basis of these observations, radiation damage in the cells can be divided into three main categories. Type I damage is oxygen independent and is responsible for loss of proliferative capacity in cells which are irradiated and maintained in the absence of oxygen until hydration is completed. Type II damage is responsible for the extra increment of reproductive death which occurs if oxygen is present during irradiation. Lastly, Type III damage arises from the presence of oxygen in the post-irradiation period. Cells irradiated and maintained in the presence of oxygen are therefore killed by a mixture of Types I, II, and III damage, while those which are irradiated in the absence of oxygen but are subsequently exposed to the gas die from a mixture of Type I and Type III damage, and so on.

The actual amount of damage of these three types can be modified greatly by treatment with H_2S and also depends on the amount of water in the cells. Exposure of the spores after irradiation to H_2S, before they are treated with oxygen, entirely abolishes all Type III damage, and H_2S treatment during irradiation reduces the amount of Type I damage. The effect of H_2S on Type II damage is uncertain. The role of water in the production of radiation damage is a little more complex (Fig. 6–3). The amount of Type I damage remains constant over a wide range of water content, though it increases by a small amount in very wet cells. Water therefore sensitizes cells to this kind of damage. The amounts of damage of Types II and III, which are very great in the driest spores, are, however, much reduced as water content is increased, and Type III damage entirely disappears in water-saturated spores. The major effect of water is therefore to protect. As a consequence the oxygen DMF can vary from as high as 10 or more in very dry spores down to the more normal 2.5 to 3.0 when the water content is higher. As in normal cells, the size of the oxygen DMF also depends on radiation quality, and the oxygen effect disappears when very high LET radiations are used.

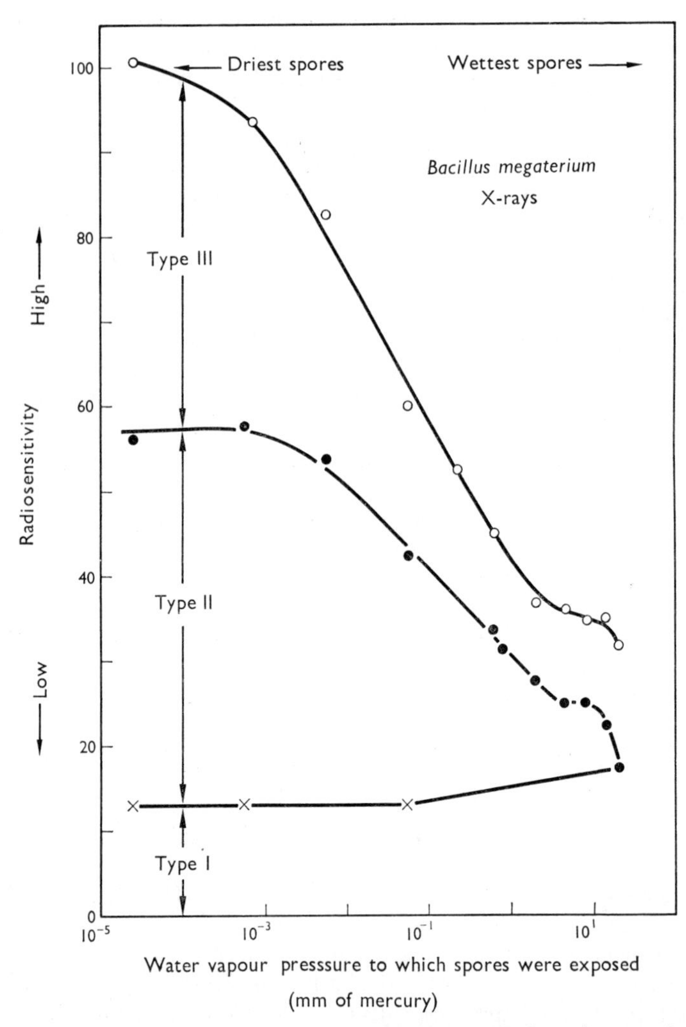

Fig. 6–3 The influence of water content on radiation damage in the bacterium *Bacillus megaterium*. (Courtesy A. Tallentire and E. L. Powers, *Radiation Research*, **20**, 270–287, and Academic Press)

Type III damage is produced by long-lived organic radicals and it is probable that Type II damage is also due to similar radicals, though in this case they are very short-lived. The chemical identity of the radicals is not known, but there are several different kinds involved in both types of damage. In the absence of oxygen these radicals are entirely harmless, but become lethal if allowed to react with oxygen. The lifetimes of the Type

III, and even Type II, radicals become very short in the presence of water, too short to allow any of them to react lethally with oxygen. Why radicals have such short lives in the presence of water is not known, though it may make them so mobile that they can quickly react with one another or with surrounding molecules of protective compounds. The observations with H_2S treatment show that prior exposure to sulphydryl compounds prevents these radicals from reacting lethally with oxygen. If RH represents a biologically active macromolecule, such as one of DNA, the reactions outlined above can be visualized as follows. Ionization within the molecule will often lead to its dissociation into a radical, with the loss of a hydrogen atom (see § 5.2).

$$RH \xrightarrow[\text{and dissociation}]{\text{ionization}} R\cdot + H^+$$

In the absence of oxygen this radical is likely to regain a hydrogen atom, perhaps directly from its surroundings or by reacting with a sulphydryl compound (XSH) or other protective agent.

$$R\cdot + H\cdot \longrightarrow RH$$
$$R\cdot + XSH \longrightarrow RH + XS\cdot$$

and if so, the biologically active molecule will be restituted. The XS• radical is presumably harmless. If oxygen is present, however, such radicals react very readily with it to form peroxy-radicals,

$$R\cdot + O_2 \longrightarrow ROO\cdot$$

which are usually unstable and dissociate in ways which lead to particularly damaging alterations in the macromolecule. It is assumed that the radicals react very readily with oxygen, at least in the absence of water, so that reactions which restore biological activity are prevented.

In cells containing more normal amounts of water, where damage can be produced by both direct and indirect action (§ 5.2), oxygen and sulphydryl compounds can modify the amount of damage in two additional ways. First, they can react with the macromolecular radicals that are produced by indirect action, in much the same way that they react with radicals produced by direct ionization, shown above. Indirect action, in which macromolecules are attacked by the radiolytic products of water, is probably responsible for the increase in Type I damage that occurs when spores have a high water content. The remaining Type I damage presumably arises from ionization directly within biologically important macromolecules. Second, oxygen and sulphydryl compounds can react with the radiolytic products of water themselves, making them innocuous in the case of sulphydryl compounds or producing new attacking species in the case of oxygen (see § 5.2), for example,

$$H\cdot + O_2 \longrightarrow HO_2\cdot$$

These secondary radicals are usually longer lived than the primary ones, and may persist to attack macromolecules.

Apart from the reactions given above, these protective and sensitizing substances may act in various other ways at the chemical stage and some, particularly high concentrations of water and perhaps cysteamine, may also have physiological effects. It is also likely that there are other protective and sensitizing substances in cells, and together they form the 'chemical environment' of DNA or other macromolecules, distributing the absorbed energy by means of a complex network of interrelated and competing chemical reactions. Sulphydryl compounds, for example, divert energy away from macromolecules and dissipate it harmlessly as chemical change within the SH compound itself. Moreover, they tend to prevent the energy absorbed in water being transmitted to vital structures. Oxygen, on the other hand, can increase the amount of energy transmitted in this way by giving rise to secondary long-lived radicals in water, and in addition can prevent energy being passed on from macromolecule to protective agent. Oxygen and sulphydryls therefore compete with one another in their reactions with macromolecules, and other modifying agents no doubt act in similar ways. The amount of damage produced in any one set of circumstances therefore depends on the particular balance struck between protecting and sensitizing reactions, which in turn will depend on the relative speeds of the different reactions, the concentrations of the protective and sensitizing agents and so on.

6.3 Repair of radiation damage

Experiments with protective and sensitizing compounds show that cellular radiosensitivity depends to a large extent on the kinds of reaction which take place during the chemical stage in the development of radiation damage (Fig. 1–1), and since these reactions are usually very fast such compounds must be present within cells during the actual period of irradiation. Radiosensitivity can also be modified, however, by changing the environment or composition of cells after irradiation, well after the end of all radiation-induced chemical reactions, and this shows that the amount of damage eventually expressed in irradiated cell populations also depends on their metabolic activities.

Although macromolecules are damaged very quickly after energy absorption, most cells that contain them appear quite normal, at least from a superficial point of view, for quite a long period after irradiation. The consequences of biochemical damage become evident only after a delay of hours, days, or even weeks, because a period of active metabolism is required before the damage is translated into visible effect. Alterations within DNA molecules, or other structures, presumably impair some aspect of the cell's metabolic capability and the injury can only be revealed if the cell attempts to carry out the impaired function; if metabolism is temporarily suppressed, by dehydrating or cooling the cell for example, the development of injury is correspondingly delayed. The molecular damage therefore undergoes 'biological amplification' during the course of metabolism, which becomes increasingly disturbed. The proportion of cells which eventually die, mutate, and so on, is not irrevocably determined at the time of irradiation, however, because cells possess the ability to *repair* at least some of their biochemical damage, and the amount of damage repaired is probably influenced by the cellular environment in the post-irradiation period.

The models, or working hypotheses, which have been developed to explain how repair occurs have concentrated on the removal of defects from DNA, not only because these macromolecules are important targets but also because their two-stranded structure seems to have been designed during the course of evolution to facilitate repair. It is less easy to conceive how other kinds of macromolecules could be repaired, though this does not of course mean that it is impossible. The repair of damaged DNA seems to require a number of enzymes, some of which use defects in the DNA as their substrate, and a supply of energy in the usual form of adenosine triphosphate (ATP). In the first stage of the repair process (Fig. 6–4), repair enzymes move along the length of the thread-like molecules of DNA, monitoring them for chemical alterations. These alterations will often locally distort the shape of the macromolecule and it is probably this distortion that repair enzymes detect. At the second stage, the defect can be corrected directly *in situ* in some cases, repairing the damage in a single step, but in other cases the defect is removed from the DNA by degrading enzymes which cut out the surrounding portion in one of the strands. The missing portion is then resynthesized by fitting nucleotides one by one into the gap. During this resynthesis it is essential to restore the original sequence of bases in the DNA strand because this sequence specifies the particular instructions or information that the DNA molecule contains. This is easily achieved, however, since only adenine–thymine and guanine–cytosine base pairs are allowed in normal DNA so that the sequence of bases in the undamaged strand opposite the gap dictates the sequence of bases which must be inserted. DNA molecules therefore contain two complementary copies of their information, which provides a safety factor and allows repair to take place. The insertion of an incorrect base would produce a mutation and it is possible that many of the radiation-induced mutations arise from such errors in repair. In diploid cells, which possess homologous pairs of chromosomes, one from the male and one from the female parent, the sequence of bases inserted into the gap may also be dictated by the sequence of bases in one of the strands in the other homologous chromosome, and this in addition allows repair to occur even when *both* strands are damaged at the same point. Diploidy therefore provides an extra safety factor, and repair of this kind has many features

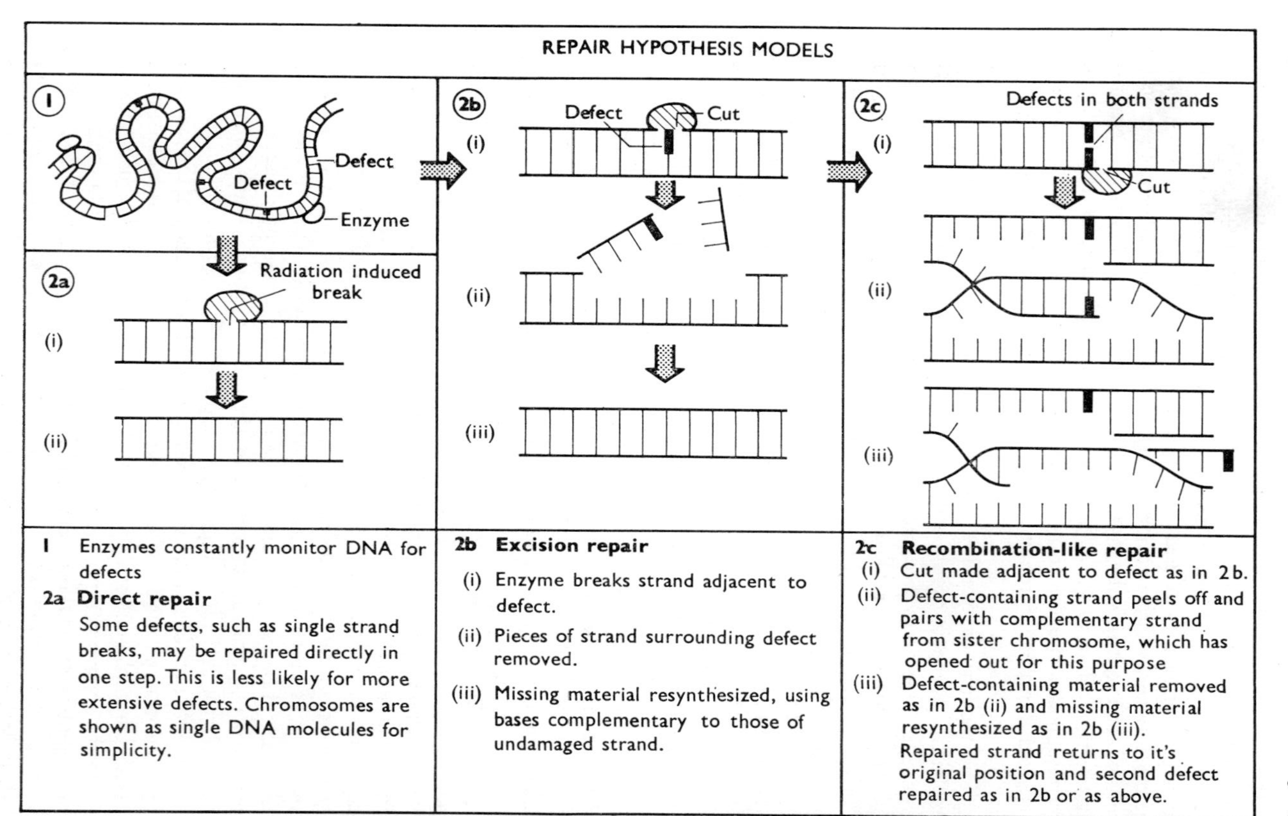

Fig. 6–4 Models for the repair of radiation-damaged DNA.

in common with genetic recombination. Evidence supporting the repair hypothesis has been obtained from a variety of experiments, but even though the models described above are likely to be correct in principle, much of the evidence is rather indirect and the details of the models have yet to be verified.

The first kind of experiment to provide evidence supporting the repair hypothesis makes use of post-irradiation treatments, such as changes in the supply of nutrients, change in the temperature of incubation, and the application of drugs which suppress different aspects of metabolism, for example oxidative phosphorylation or the synthesis of protein or nucleic acids. These treatments influence the metabolic activities of cells in the period after irradiation, and probably either increase or decrease the amount of biochemical damage that is repaired. Because the evidence is indirect, however, it is possible that in some cases the treatments merely modify the expression of the biochemical damage rather than its repair, a variable proportion being bypassed or suppressed in different conditions. They have been found to be effective with a wide range of organisms or types of cell, and can modify the proportion of cells losing their proliferative capacity (e.g. ALPER, 1961; PATRICK and HAYNES, 1964; PHILLIPS and TOLMACH, 1966) as well as the frequency of induced chromosome aberrations (BEATTY and BEATTY, 1959; KIHLMAN and HARTLEY, 1968) or gene mutations (KADA *et al.*, 1961; KIMBALL, 1963).

These post-irradiation treatments are thought to act in two main ways. Some, particularly the rather non-specific ones like changes in the supply of nutrients, probably act indirectly by increasing or decreasing the time available for repair. Defects in DNA molecules are likely to be particularly damaging to the cell at certain critical stages in their growth cycle, such as when new DNA is synthesized or the cell divides, and since some repair processes are rather slow, rapidly growing cells will be unable to repair all the defects before they enter the critical stage. If growth can be slowed down, by limiting the supply of nutrients and omitting inessential substances, or even temporarily stopped without loss of viability, by suspending the cells in buffer or saline, a much greater proportion of the defects can be repaired. Other post-irradiation treatments probably act more directly on repair processes. For example, since nucleotides are required to patch up the gaps produced during excision repair (Fig. 6–4), drugs which prevent their synthesis are also likely to prevent repair.

Many post-irradiation treatments are non dose-modifying and have a greater effect after low than high doses, possibly because the repair process itself is damaged by the radiation or the damage produced by high doses is less easily repaired. In the case of survival curves, post-irradiation treatments usually have greater effect on the 'shoulder' than on the final slope, which suggests that the size of the shoulder reflects the capacity to repair damage as well as multihit or multitarget inactivation. In many cases such treatments are also less effective when cells have been exposed to densely ionizing radiations, probably because the damage produced by these radiations is less easy to repair.

The second type of experiment which provides support for the repair hypothesis depends on splitting the dose of radiation into two or more increments, separated by varying intervals of time, or on varying the dose rate by giving the same total dose over a shorter, or longer period. Cells often suffer less damage when the radiation dose is split into several increments or delivered at a low dose rate and this 'sparing' effect has been ascribed to the repair of biochemical damage in the interval between dose increments or during the actual irradiation period when the dose rate is low.

Elkind and his colleagues have made a detailed study of this 'sparing' effect with regard to loss of proliferative capacity in cultured mammalian cells (reviewed in ELKIND and WHITMORE, 1967) and the result of one of their experiments is shown in Fig. 6–5. If these cells receive a single dose of 992 rads of X-rays at the beginning of the experiment only 0.186 per cent of them remain capable of forming a colony but if this dose is split into two increments, an initial 505 rads followed at varying intervals by 487 rads, a higher proportion of cells survive. This proportion rapidly increases to about 0.55 per cent when the interval between increments is 1–2 hr (Fig. 6–5, open circles), and further variation is probably due to changes in the inherent radio sensitivity of cells cultured for more than 2 hr rather than changes in repair. The shape of the survival curve when unsplit doses are used (Fig. 6–5, filled circles) corresponds closely to that expected from a multihit model of inactivation and if this model is appropriate many cells which eventually survive must contain *sub-lethal* damage at the time of irradiation, particularly if high doses have been used. If, for example, two hits are required to kill a cell, many that survive will contain targets which have been sub-lethally damaged by one hit. In this circumstance, cells which have received the first of two dose increments will mostly be of two types; those which have received two or more hits, inevitably destined to die, and those which have received one hit. Cells which have received no hits will be very rare, except at very low doses. If the second dose increment follows quickly after the first, many cells will therefore die from a *single* extra hit because they have already been hit once before. If, however, a longer period elapses between the two dose increments, during which sub-lethal damage is repaired, such cells will die only if they receive *two* hits in the second increment of dose, and those receiving only one hit will now survive. If all sub-lethal damage is repaired, survival after two dose increments will simply be the product of the surviving fractions from each dose increment alone. Since 8.2 per cent of the cells survive a dose of 505 rads, and 9.5 per cent a dose of 487 rads (Fig. 6–5), full repair should allow 0.78 per cent (=8.2 per cent × 9.5 per cent) of the cells to survive. An interval of 1–2 hr between dose increments, which raises survival to 0.55 per cent, therefore allows the repair of much, but not all, sub-lethal damage.

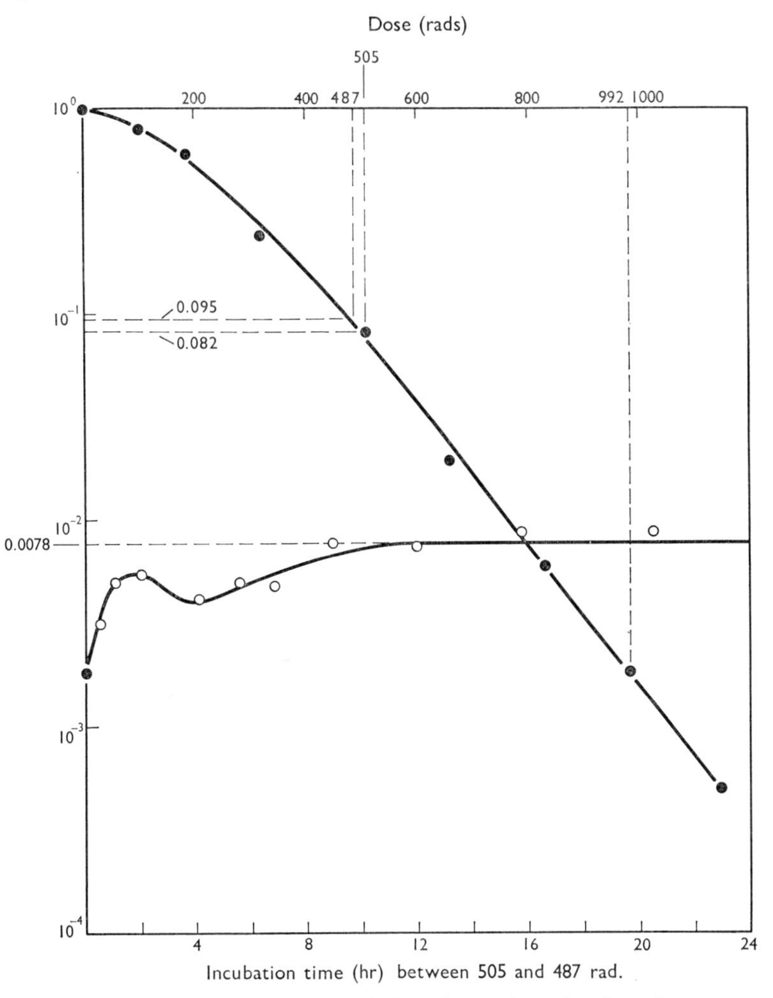

Fig. 6–5 Survival curve (closed circles) and two-dose fractionation curve (open circles) of cultured Chinese hamster cells (X-rays). (Courtesy M. M. Elkind and H. Sutton, *Radiation Research*, **13**, 556–593, and Academic Press)

The 'sparing' effect of dose splitting is greatest when sparsely ionizing radiations are used, and decreases as ion density increases, often disappearing altogether at very high ion densities. This decrease parallels the decrease in size of 'shoulder' on survival curves, and cell types, conditions, or radiations which produce 'shoulderless' survival curves also fail to demonstrate 'sparing' effects. The 'sparing' effect can be suppressed, in conditions where it is normally found, by treating the irradiated cell population in various ways between successive dose increments and this emphasizes the physiological nature of the 'sparing' effect. Survival can be enhanced by reducing the rate at which the radiation dose is delivered as well as by dose splitting, probably because more damage can be repaired when the dose rate is low, and this enhancement also diminishes as the radiation ion density is increased.

Split-dose and dose-rate experiments have also been used to examine the formation of chromosome aberrations, particularly chromosome and chromated exchanges, and these types of experiment were in fact first used for this purpose (see reviews by LEA, 1946 and EVANS, 1968). These exchange aberrations can only occur if two chromosomes or chromatids are damaged and if the two damaged sites lie close together (see Chapter 4). When sparsely ionizing radiations are used, each of these sites is likely to have been damaged by a separate track of ionizations since the chance that a single track will damage both is very small (see § 3.4), so the yield of exchange aberrations is proportional to the square of the dose; that is, halving the dose reduces the yield of exchanges to a quarter. When a dose of radiation is split into two equal parts, the second following quickly after the first, the final yield of exchanges will therefore have arisen in three ways. One-quarter will be due to two hits received during the first dose increment, one-quarter to two hits received during the second increment, and one-half to one hit received during the first increment combined with one hit received during the second. This is, of course, a situation closely analogous to the interpretation placed on Elkind's split-dose experiment, described above. When the two dose increments are separated by an increasing time interval, the yield of exchanges drops and finally reaches a value one-half the size found with the unsplit dose. This is because the third group of exchanges, those due to single hits in each dose increment, becomes smaller and finally disappears. During the interval between the two doses, therefore, the initial radiation damage is repaired or, more precisely, is changed in some way such that it is no longer available to take part in an exchange; it is not necessary to assume that exchange takes place quickly, only that it must be initiated in some way during this period. The process takes about twenty minutes in *Tradescantia* pollen grains, and generally takes between ten minutes and a few hours depending on materials and conditions.

The period during which initial chromosome damage remains available for exchange can be lengthened considerably by treating irradiated cells with various drugs, such as those which inhibit the synthesis of ATP or protein (WOLFF, 1959), during the interval between dose increments, and this demonstrates the physiological nature of the repair or modification process. As with the repair of sub-lethal damage, the effect of splitting the dose disappears when densely ionizing radiations are used. The yield of exchanges becomes linearly related to dose in this case, probably because a single densely ionizing track can damage both sites needed for an exchange. The yield of exchanges is also much reduced at low dose rates, so that loss of proliferative capacity and the formation of chromosome aberrations respond in closely similar, though not identical ways to changes of irradiation regime.

Finally, the last of the main types of experiment which provide support for the repair hypothesis are those which make use of radiosensitive mutants. If repair depends on the activities of special enzymes, the structure of these enzymes will be determined by genes located somewhere in the genetic material in exactly the same way as for enzymes concerned with other metabolic activities. Cells which are mutant with respect to one of these genes will lack that particular enzyme, and as a consequence will become much more radiosensitive because they can no longer repair some of the radiation damage. A number of very radiosensitive mutants are known in bacteria and yeast, owing to mutation in one of about five different genes, and several are known in other organisms. This suggests that there are at least five different enzymes or proteins concerned in the repair of radiation damage and there are probably more. The properties of these radiosensitive mutants provide useful information about repair activities.

Comparisons between the survival curves of normal and mutant cells show that absence of repair activity is rarely dose-modifying, because the 'shoulder' is usually much reduced or absent. As might be expected, the mutants are usually less susceptible to the modifying action of post-irradiation treatments and in bacteria they often exhibit smaller oxygen dose-modifying factors. ALPER (1963) has therefore suggested that oxygen-dependent radiation damage is much less amenable to repair than oxygen-independent damage. The damage produced by densely ionizing radiations is also much less amenable to repair because the difference between the survival curves of normal and mutant cells is much less, and may disappear, when high LET radiations are employed. Apart from their increased radiosensitivity, a number of mutants are also unable to carry out normal genetic recombination, which suggests that some radiation damage is repaired by a recombination-like process (Fig. 6–5). Mutations of any kind can be induced by radiation only at much lower frequencies than normal in these recombination deficient mutants. That is, radiation can produce only very few mutations anywhere in the genetic material when recombination is suppressed. It is therefore likely that many of the mutations induced by radiation in normal cells arise because of errors during the recombination-like repair of damaged DNA, though some no doubt arise in other ways.

When viewed as a whole, the combined evidence from many experiments provides a considerable amount of circumstantial support for the repair hypothesis but direct confirmation can only be obtained when the various repair enzymes have been isolated, purified, and the details of their activities studied in the test-tube. Although this has not yet been done for enzymes concerned in the repair of damage produced by ionizing radiations, some which repair defects in DNA caused by ultra-violet radiations have been studied in this way, and the work shows that repair enzymes do in fact exist. We are therefore faced with the problem of explaining why many, and perhaps all, organisms possess systems of repair enzymes, since few can have been exposed to large doses of ionizing radiations, or even short-wave ultra-violet light, in normal circumstances. What, then, is the evolutionary significance of repair systems? The answer most probably is that they arose, during the course of evolution, to remove defects which arise spontaneously in the DNA of cells growing in normal circumstances. The genetic material is maintained and transmitted with amazing fidelity, and a single gene may be duplicated and passed on a million or more times without error. Although DNA molecules have a fairly stable structure it seems likely that such fidelity can only be maintained by active removal of defects as they arise. Apart from lethal defects in DNA, some may give rise to mutations and most mutations are deleterious. Nevertheless some mutations are essential, to allow evolutionary change to take place, and repair enzymes may regulate the mutation rate and maintain it at an optimum level. Finally, some enzymes which perhaps first arose to repair DNA defects, are now also involved in crossing-over and genetic recombination, and the reshuffling of genes to produce new combinations has been a most important factor in the evolution of the majority of organisms.

References

ALPER, T. (1961). 'Effects on Subcellular Units and Free-living Cells'. In *Mechanisms in Radiobiology, Vol. 1*, Ed. M. ERRERA and A. FORSSBERG, Academic Press, New York and London.

ALPER, T. (1963). *Nature, Lond.*, **200**, 534–536.

BACQ, Z. M. and ALEXANDER, P. (1961). *Fundamentals of Radiobiology*, 2nd Edn, Pergamon Press, Oxford and London.

BEATTY, A. V. and BEATTY, J. W. (1959). *Am. J. Bot.*, **46**, 317–323.

BREWEN, J. G. and BROCK, R. D. (1968). *Mutation Res.*, **6**, 245–255.

COLE, A. (1965). 'The Study of Radiosensitive Structures with Low Voltage Electron Beams'. In *Cellular Radiation Biology*. The Williams and Wilkins Company, Baltimore.

CONGER, A. D. (1967). *Mutation Res.*, **4**, 449–459.

DAVIES, D. R. and EVANS, H. J. (1966). 'The Role of Genetic Damage in Radiation-Induced Cell Lethality'. In *Advances in Radiation Biology, Vol. 2*, Academic Press, New York.

ELKIND, M. M. and WHITMORE, G. F. (1967). *The Radiobiology of Cultured Mammalian Cells.* Gordon and Breach, New York, London, Paris.

EVANS, H. J. (1968). 'Repair and Recovery at Chromosome and Cellular Levels: Similarities and Differences'. In *Brookhaven Symposia in Biology*, **20**.

ERIKSON, R. L. and SZYBALSKI, W. (1963). *Radiat. Res.*, **20**, 252–262.

FREIFELDER, D. (1968a). *J. molec. Biol.*, **35**, 303–309.

FREIFELDER, D. (1968b). *Virology*, **36**, 613–619.

GINOZA, W. (1967). 'The Effect of Ionizing Radiation on Nucleic Acids of Bacteriophages and Bacterial Cells'. In *A. Rev. nucl. Sci.*, **17**.

HEDDLE, J. A. and BODYCOTE, D. J. (1970). *Mutation Res.*, **9**, 117–126.

HOWARD-FLANDERS, P. and MOORE, D. (1958). *Radiat. Res.*, **9**, 422–437.

KADA, T., DOUDNEY, C. O. and HAAS, F. L. (1961). *Genetics*, **46**, 683–702.

KIHLMAN, B. A. and HARTLEY, B. (1968). *Hereditas*, **59**, 439–463.

KIMBALL, R. F. (1963). 'The Relation of Repair to Differential Radiosensitivity in the Production of Mutations in *Paramecium*'. In *Repair from Genetic Radiation Damage*, Ed. F. H. SOBELS, Pergamon Press, Oxford and London.

LEA, D. E. (1946). *Actions of Radiations on Living Cells*, Cambridge University Press.

MCGRATH, R. A. and WILLIAMS, R. W. (1966). *Nature Lond.*, **212**, 534–535.

MORTIMER, R., BRUSTAD, T., and CORMACK, D. V. (1965). *Radiat. Res.*, **26**, 465–482.

NEARY, G. J. (1965). *Int. J. Radiat. Biol.*, **9**, 477–502.

PATRICK, M. H. and HAYNES, R. H. (1964). *Radiat. Res.*, **23**, 564–579.

PHILLIPS, R. A. and TOLMACH, L. J. (1966). *Radiat. Res.*, **29**, 413–432.

POWERS, E. L. (1965). 'Some Physicochemical Bases of Radiation Sensitivity in Cells'. In *Cellular Radiation Biology*. The Williams and Wilkins Co., Baltimore.

REVELL, S. H. (1955). 'A New Hypothesis for Chromatid Exchange'. In *Proceedings of the Radiobiology Symposium*, Liège, 1954, Butterworth, London.

SAX, K. (1940). *Genetics*, **25**, 41–68.

DE SERRES, F. J., WEBBER, B. B., and LYMAN, J. T. (1967). *Radiat. Res. Suppl.* **7**, 160–171.

SKARSGARD, L. D., KIHLMAN, B. A., PARKER, L., PUJARA, C. M., and RICHARDSON, S. (1967). *Radiat. Res. Suppl.*, **7**, 208–221.

SPARROW, A. H. (1965). 'Relationship between Chromosome Volume and Radiation Sensitivity in Plant Cells'. In *Cellular Radiation Biology*. The Williams and Wilkins Co., Baltimore.

TODD, P. (1967). *Radiat. Res., Suppl.*, **7**, 196–207.

ULRICH, H. (1951). *Naturwissenschaften*, **38**, 530–543.

VON BORSTEL, R. C. and ROGERS, W. (1958). *Radiat. Res.*, **8**, 248–253.

WOLFF, S. (1959). *Radiat. Res., Suppl.*, **1**, 453–462.

ZIMMER, K. G. (1961). *Studies on Quantitative Radiation Biology*, Oliver & Boyd, Edinburgh and London.

Progress Papers 3

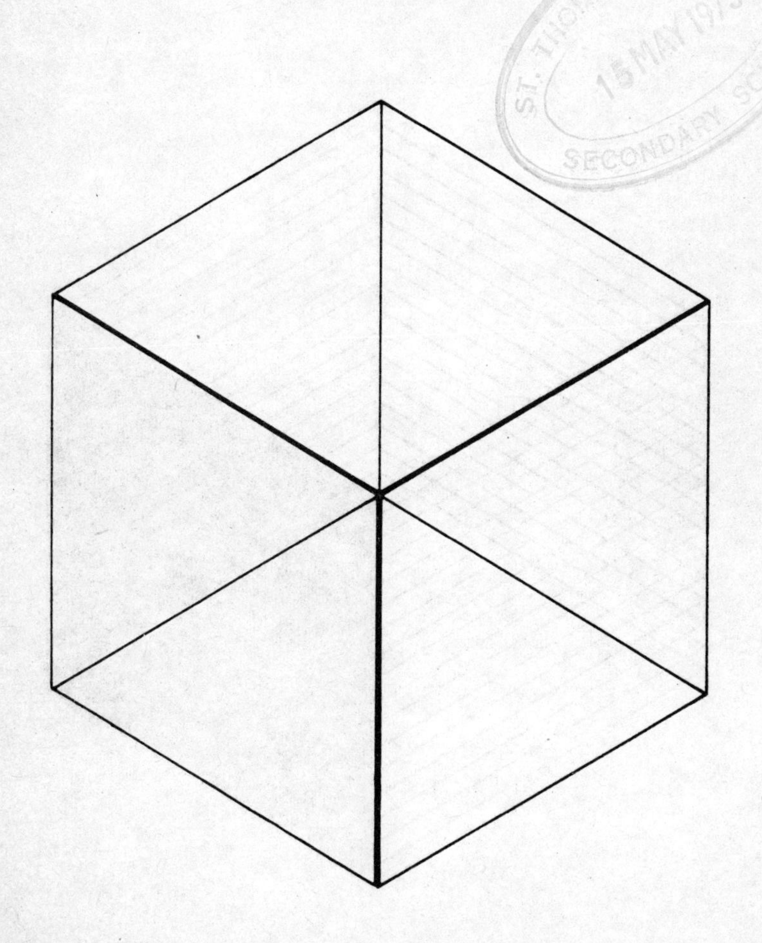

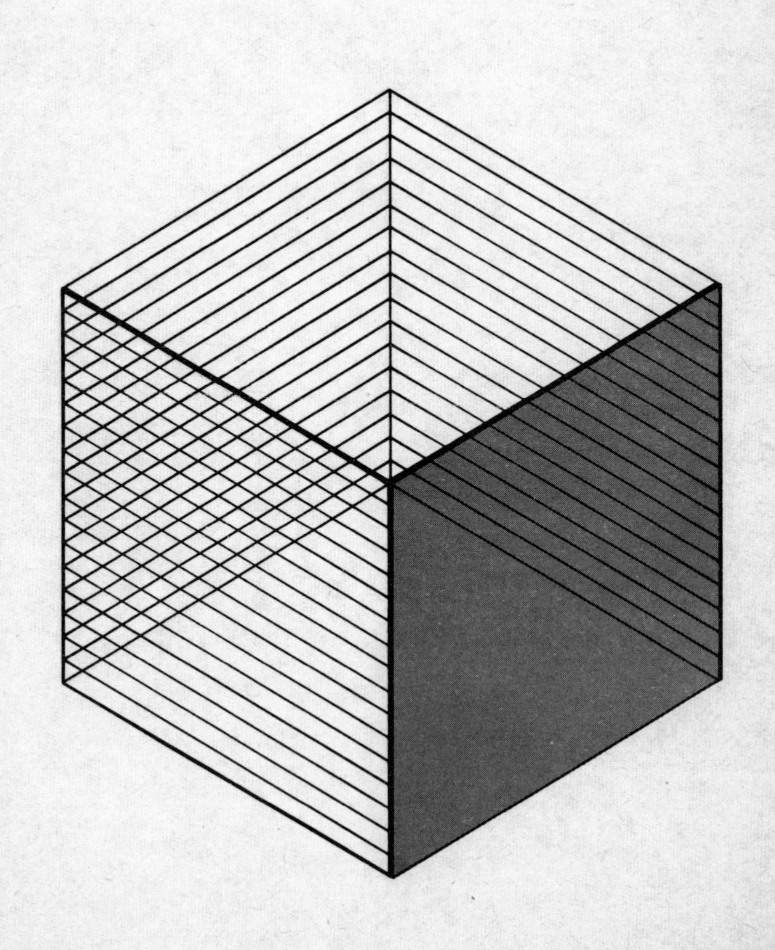

Progress Papers 3

Second Edition

Scottish Mathematics Group

Chambers

Blackie & Son Limited
Bishopbriggs · Glasgow
5 Fitzhardinge Street · London W1

W & R Chambers Limited
11 Thistle Street · Edinburgh 2
6 Dean Street · London W1

First Published 1972

Designed by James W. Murray

International Standard Book Numbers
Blackie 216.89410.7
Chambers 550.75943.3

Printed in Great Britain by
Butler & Tanner Limited · Frome and London
Set in 10 on 12 pt Monotype Times Roman

Members associated with this book

W. T. Blackburn
Dundee College of Education

Brenda I. Briggs
Formerly of Mary Erskine School for Girls

W. Brodie
Trinity Academy

C. Clark
Formerly of Lenzie Academy

D. Donald
Formerly of Robert Gordon's College

R. A. Finlayson
Allan Glen's School

Elizabeth K. Henderson
Westbourne School for Girls

J. L. Hodge
Madras College

J. Hunter
University of Glasgow

T. K. McIntyre
High School of Stirling

R. McKendrick
Langside College

W. More
Formerly of High School of Dundee

Helen C. Murdoch
Hutchesons' Girls' Grammar School

A. G. Robertson
John Neilson High School

A. G. Sillitto
Formerly of Jordanhill College of Education

A. A. Sturrock
Grove Academy

Rev. J. Taylor
St. Aloysius' College

E. B. C. Thornton
Bishop Otter College

J. A. Walker
Dollar Academy

P. Whyte
Hutchesons' Boys' Grammar School

H. S. Wylie
Govan High School

Only with the arithmetic section

R. D. Walton
Dumfries Academy

This series of booklets of Progress Papers is based on the content of the textbooks *Modern Mathematics for Schools* (Second Edition). Each booklet relates to the textbook of corresponding number.

The Aims of the Series

The main aims of the booklets and associated material are

(*a*) to encourage pupils in their progress through the mathematics course, and
(*b*) to provide teachers with a quick and appropriate check on the pupils' attainments.

The tests can be used to obtain a *continuous* assessment of progress, although from time to time there will be a need to set questions which require a more sustained form of answer.

The Material

In this booklet there are four tests on the content of each chapter of the corresponding textbook, except in the case of the chapters on *Probability*, and on *Time*, *Distance*, *Speed*, for which there is one ***A*** test and one ***B*** test. The tests headed ***A1*** and ***A2*** are based on the ***A*** set of Exercises in the textbook; those headed ***B1*** and ***B2*** are more difficult and are based on the ***B*** set of Exercises.

In order to simplify the teacher's task as far as possible, pupils' answer sheets are available, and also a teacher's marking overlay book. In this way each sheet can be marked and totalled quickly and accurately. It may indeed be appropriate from time to time for pupils to mark their own tests so that an element of learning can be incorporated in the process.

In brief, the system requires that the teacher should possess

(*a*) a class set of booklets to be issued to the class for each test;
(*b*) the relevant book of marking overlays;
(*c*) pupils' answer sheets, which are expendable after the scores have been recorded.

The Method of Use

Each test consists of five questions requiring ***True/False/Don't know*** responses, followed by ten questions of a multiple-choice type with five responses, except in a few cases where the situation seems to justify only four responses.

The purpose of including the ***E*** response throughout is two-fold: in the first place the ***E*** response avoids a pupil feeling obliged to circle ***A, B, C*** or ***D*** as a guess; secondly, ***E*** responses may indicate to the teacher those aspects of the course in which the pupils are insecure. Where a question does not justify more than four responses, these will be lettered ***A, B, C, E***; the answer sheet will still have ***A, B, C, D, E*** on every line.

It is recommended that 1 or 0 be awarded for each answer in the 1–5 set, and 2 or 0 for each answer in the 6–15 set, giving a maximum mark of 25. It is not considered necessary to award negative scores for incorrect answers.

All the questions have been pre-tested in schools, and an analysis of the results has been made in order to ensure that the content of the questions and the nature of the distractors are reasonable and appropriate, and that the overall scores will provide encouragement for most pupils. It was found that the majority of pupils answered a test in 10 to 20 minutes, depending on their ability and on the difficulty of the test.

Pupils should be instructed not to write on the *Progress Papers*, which are intended to be withdrawn after each test. Any necessary working should be permitted on jotters or on scrap paper, and in one or two cases squared paper is required.

Notation for sets of numbers used in these Papers

The set of natural numbers. $N = \{1, 2, 3, \ldots\}$
The set of whole numbers. $W = \{0, 1, 2, \ldots\}$
The set of integers. $Z = \{\ldots, -2, -1, 0, 1, 2, \ldots\}$
The set of rational numbers. Q
The set of real numbers. R
The set of prime numbers. $\{2, 3, 5, 7, 11, \ldots\}$

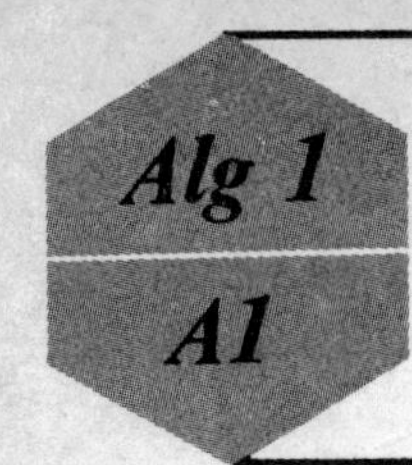

Relations, Mappings and Graphs

Questions 1–5: On the answer sheet, circle T for True, F for False, or E for 'don't know'.

---

1 The arrow diagram in Figure 1 shows the relation *is greater than* from set P to set Q.

2 Figure 1 shows a mapping from set P to set Q.

3 Figure 2 indicates that no boy in set R is 17 years old.

4 The only boy in set R shown in Figure 2 who is aged 15 is Bob.

5 The set of ordered pairs $\{(1, 4), (2, 8), (3, 12)\}$ describes the relation *is quarter of* from set $\{1, 2, 3\}$ to set $\{4, 8, 12\}$.

Questions 6–15: On the answer sheet, circle A, or B, or C, or D, or E (don't know).

---

6 From the arrow diagram in Figure 3, the girl who went on holiday to two countries is
A Ann *B* Jean *C* Mary *D* Jane *E*

7 The relation from set X to set Y in Figure 4 is
A *is one more than* *B* *is one less than* *C* *is double*
D *is half* *E*

Questions 8 and 9 refer to Figure 5, which is the graph of a relation from P to Q.

8 The relation can be described by
A $\{a, b, c\}$ *B* $\{(y, a), (y, b), (y, c)\}$
C $\{(a, y), (b, y), (c, y)\}$ *D* $\{(a, x), (b, y), (c, z)\}$ *E*

9 The members of the first set are
A $\{a, b, c\}$ *B* $\{x, y, z\}$ *C* $\{y\}$ *D* $\{(a, y), (b, y), (c, y)\}$ *E*

10 A relation x *is greater than* y is defined on the set $\{5, 6, 8, 9\}$. A member of this relation is
A $(5, 6)$ *B* $(6, 9)$ *C* $(6, 6)$ *D* $(6, 5)$ *E*

11 The graph in Figure 6 shows the relation on the given sets

A $x < y$ ***B*** $x = y$ ***C*** $x > y$ ***E***

12 A relation on the set $\{a, b, c, d, e, i, k\}$ is defined by the open sentence *x and y are vowels*. A member of this set is

A (a, b) ***B*** (b, b) ***C*** (e, i) ***D*** (c, e) ***E***

13 $P = \{4, 5, 6\}$ and $Q = \{5, 6\}$. If $x \in P$ and $y \in Q$, then the relation *x is greater than y* is given by

A $\{(4, 6)\}$ ***B*** $\{(5, 6)\}$ ***C*** $\{(6, 5)\}$ ***D*** $\{(6, 6)\}$ ***E***

14 $S = \{a, e, i, o\}$ and $T = \{n, r\}$. If $x \in S$ and $y \in T$, then one of the ordered pairs in the relation *y followed by x spells a two-lettered English word* is

A (n, o) ***B*** (n, e) ***C*** (a, n) ***D*** (e, n) ***E***

15 $W = \{0, 1, 4, 9\}$, $X = \{1, 4, 9\}$, $Y = \{a, z, b, k\}$, $Z = \{0, 1, 2, 3, 4\}$.

Which pair of these sets can be put in one-to-one correspondence?

A W and X ***B*** X and Y ***C*** W and Y

D X and Z ***E***

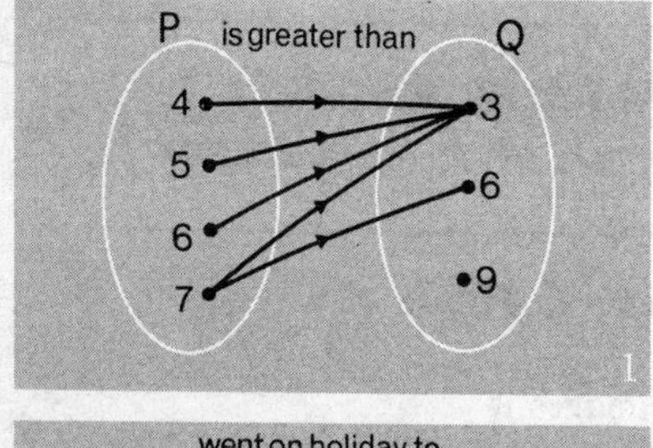

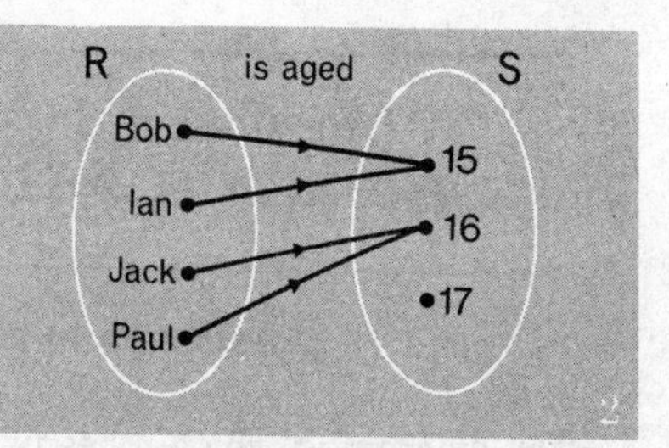

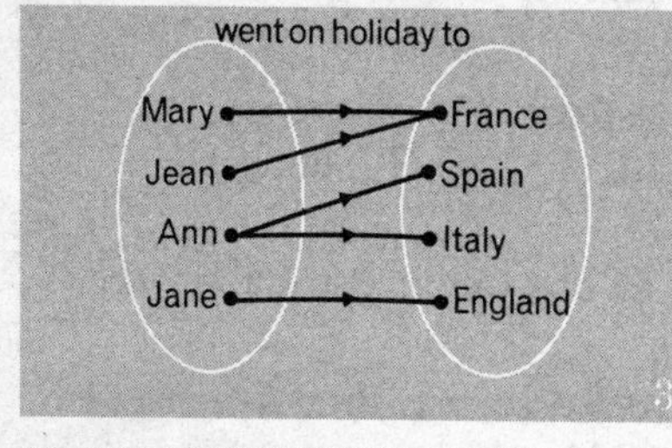

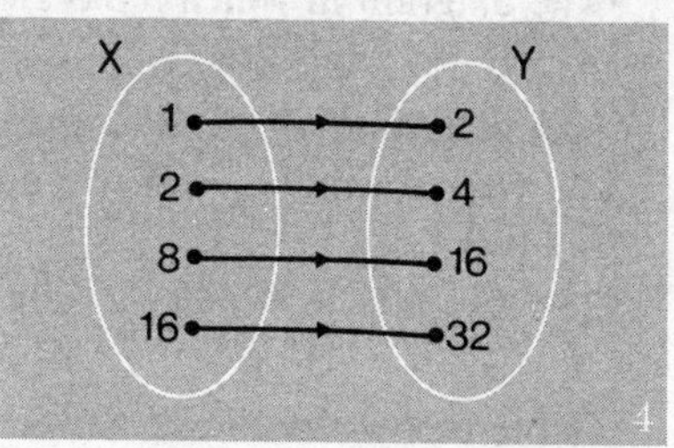

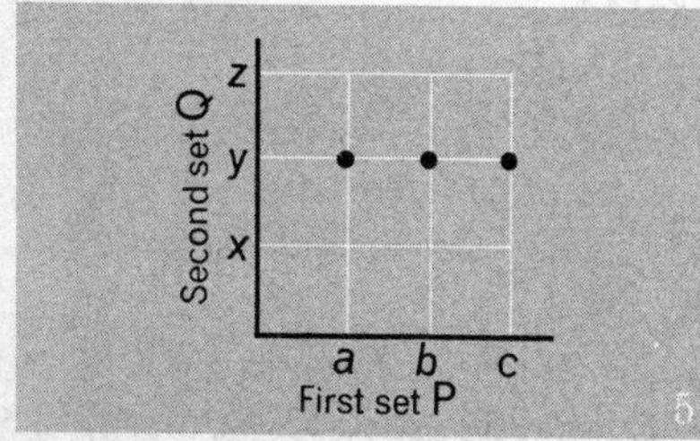

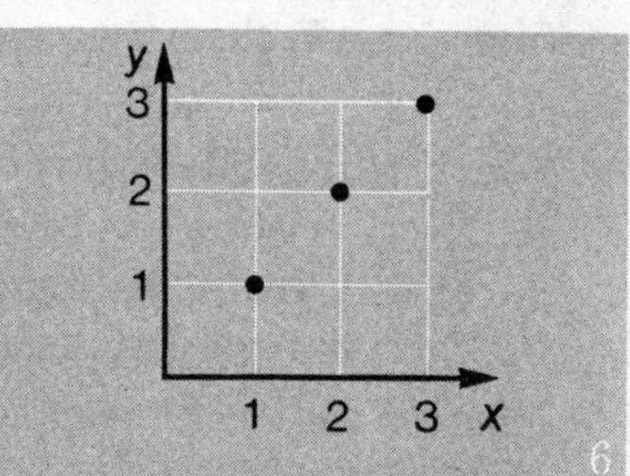

Alg 1

A2

Relations, Mappings and Graphs

Questions 1–5: On the answer sheet, circle T for True, F for False, or E for 'don't know'.

---

1 The arrow diagram in Figure 1 shows a relation from set P to set Q.

2 Figure 1 shows a mapping from set P to set Q.

3 Figure 2 shows a mapping from set B to set A.

4 The sets shown in Figure 3 are in one-to-one correspondence.

5 If $S = \{4, 2, 1\}$ and $T = \{a, b\}$, then the set of ordered pairs $\{(4, a), (2, a), (2, b), (1, b)\}$ defines a mapping from S to T.

Questions 6–15: On the answer sheet, circle A, or B, or C, or D, or E (don't know).

Questions 6 to 9 refer to Figure 4.

6 The diagram in which p has two images is
A (i) *B* (ii) *C* (iii) *D* (iv) *E*

7 The diagrams which show mappings from X to Y are
A (i) and (ii) *B* (ii) and (iii) *C* (ii) and (iv)
D (iii) and (iv) *E*

8 The diagram in which the sets are in one-to-one correspondence is
A (i) *B* (ii) *C* (iii) *D* (iv) *E*

9 If the arrows in Figure 4 were reversed so that the relations were from Y to X, the only mapping would be
A (i) *B* (ii) *C* (iii) *D* (iv) *E*

10 Which of the following is false?
A Every relation can be given by a set of ordered pairs.
B Every relation has sense. *C* Every relation is a mapping.
D Every mapping is a relation. *E*

11 Figure 5 shows the mapping from set M to set N given by
A $\{1, 2, 3, 4\}$ *B* $\{2, 3, 4, 5\}$ *C* $\{(1, 2), (2, 3), (3, 4), (4, 5)\}$
D $\{(2, 1), (3, 2), (4, 3), (5, 4)\}$ *E*

12 The mapping shown in Figure 5 can be given by

A $n \rightarrow n$ ***B*** $n \rightarrow n+1$ ***C*** $n \rightarrow 2$ ***D*** $n \rightarrow 5$ ***E***

13 How many mappings are possible from set $\{m\}$ to set $\{r, s\}$?

A 1 ***B*** 2 ***C*** 3 ***D*** 4 ***E***

14 A mapping on the set {0, 1, 2, 3, 4, 6} is defined by $n \rightarrow \frac{1}{2}(n-1)$ if n is odd, and $n \rightarrow 2$ if n is even. The mapping is

A {0, 1, 2} ***B*** {0, 1, 2, 3, 4, 6}

C {(0, 2), (1, 0), (2, 2), (3, 1), (4, 2), (6, 2)}

D {(2, 0), (0, 1), (2, 2), (1, 3), (2, 4), (2, 6)} ***E***

15 $K = \{2, 3, 4, 5, 6\}$. The relation *is 2 less than* on the set K is shown in an arrow diagram. The number of arrows is

A 2 ***B*** 3 ***C*** 4 ***D*** 5 ***E***

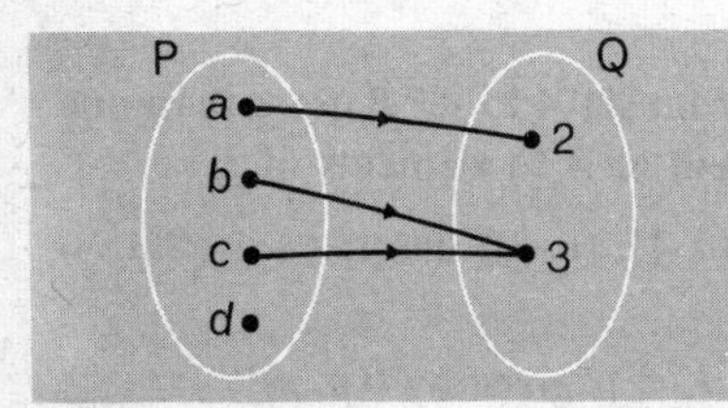

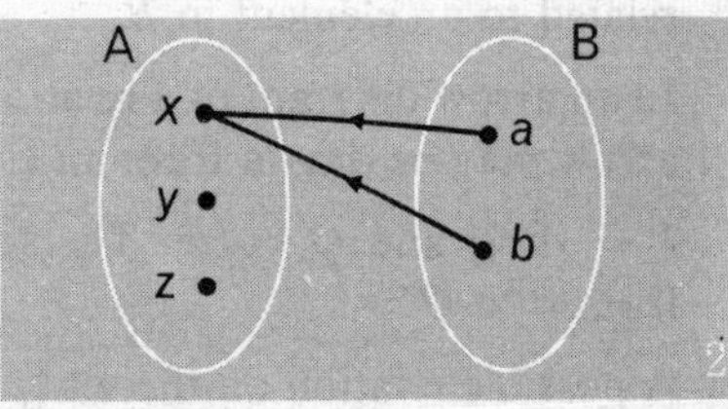

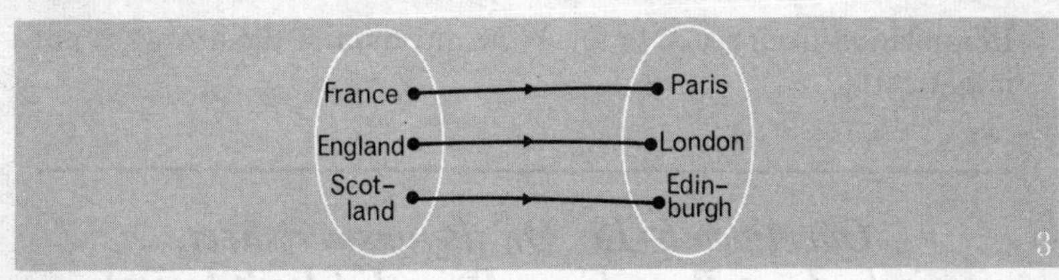

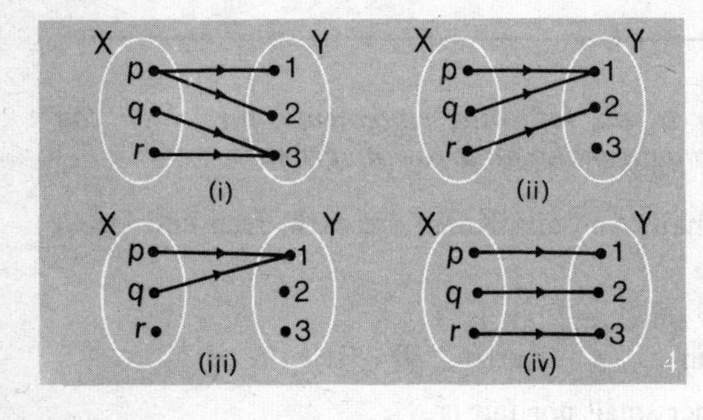

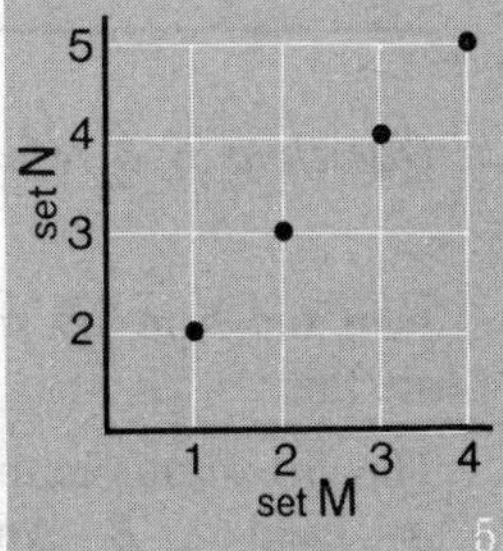

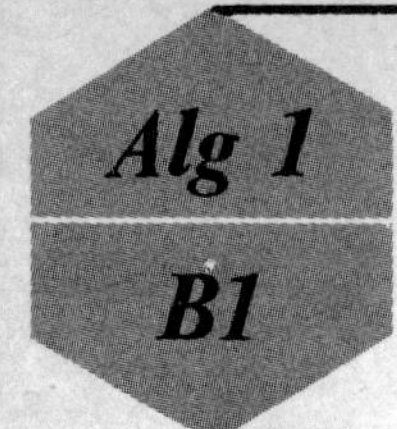

Relations, Mappings and Graphs

Questions 1–5: On the answer sheet, circle T for True, F for False, or E for 'don't know'.

---

1 In a mapping from a set A to a set B every element in A must be related to an element in B.

2 In an arrow diagram of a relation from set A to set B more than one arrow may be drawn from an element of A to elements of B.

3 $P = \{1, 2\}$ and $Q = \{a\}$. The set $\{(a, 1), (a, 2)\}$ describes a relation from P to Q.

4 A possible relation on the set $\{p, q, r, s\}$ is given by $\{(p, p)\}$.

5 In a relation from set X to set Y the direction of the arrows is not important.

Questions 6–15: On the answer sheet, circle A, or B, or C, or D, or E (don't know).

Questions 6–8 refer to the following statements, for which you might make an arrow diagram.

Jean and Ann are small. Ann and Karen are dark. Jean and Helen are fair.

6 The girl who is both small and fair is
A Jean *B* Ann *C* Karen *D* Helen *E*

7 The girl who is neither small nor fair is
A Jean *B* Ann *C* Karen *D* Helen *E*

8 The girl who is fair but not small is
A Jean *B* Ann *C* Karen *D* Helen *E*

9 Mrs Smith has two children, and Mrs Brown has three children. In an arrow diagram showing the relation *is the child of*, how many arrows are there?

A 1 ***B*** 2 ***C*** 3 ***D*** 5 ***E***

10 A relation from set P to set Q is given by $\{(2, 1), (8, 4)\}$. Q contains the subset

A $\{1, 2, 4, 8\}$ ***B*** $\{2, 8\}$ ***C*** $\{1, 4\}$ ***D*** $\{4, 8\}$ ***E***

11 The relation from set P to set Q in question *10* is

A *is half* ***B*** *is 1 more than* ***C*** *is the square of*
D *is double* ***E***

Questions 12 and 13 refer to Figure 1 for a mapping from set M to set N.

12 The image of 1 is

A 0 ***B*** 1 ***C*** 2 ***D*** 3 ***E***

13 The relation from set M to set N is

A *is 1 more than twice* ***B*** *is the square of*
C *is 1 more than the square of* ***D*** *is 1 less than the square of* ***E***

14 Set P consists of the factors of 4, set Q the factors of 8, set R the factors of 10, set S the factors of 12. Which sets can be put in one-to-one correspondence?

A P and Q ***B*** Q and R ***C*** R and S ***D*** P and S ***E***

15 The relation *is the square of* on the set $\{-1, 0, 1, 4\}$ is given by

A $\{(1, -1), (0, 0), (1, 1)\}$ ***B*** $\{(-1, 1), (0, 0), (1, 1)\}$
C $\{(-1, -1), (0, 0), (1, 1), (4, 2)\}$
D $\{(-1, 1), (0, 0), (1, 1), (16, 4)\}$ ***E***

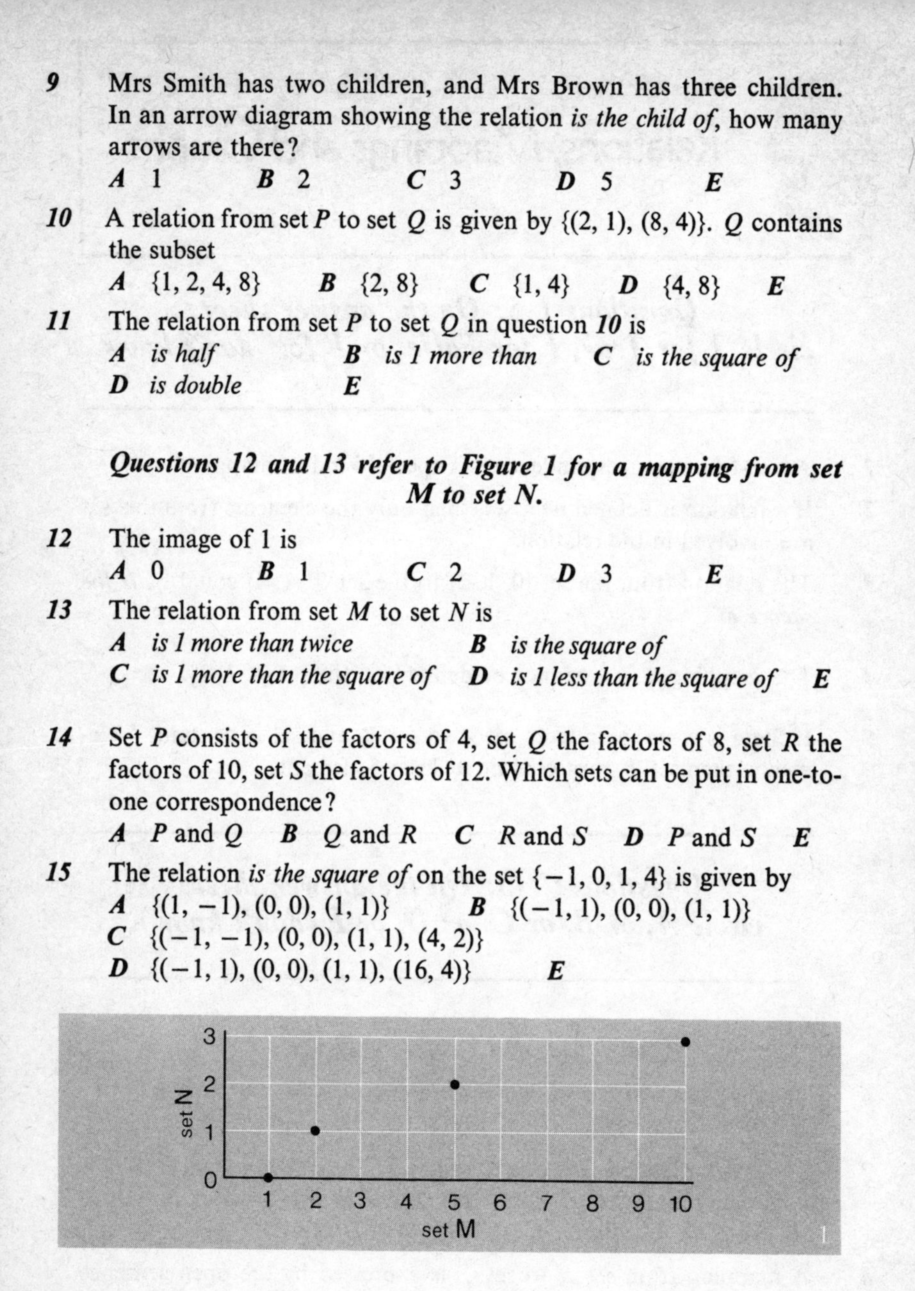

Alg 1

B2

Relations, Mappings and Graphs

Questions 1–5: On the answer sheet, circle T for True, F for False, or E for 'don't know'.

---

1 A one-to-one correspondence is a special kind of mapping.

2 If a relation is defined *on* a set, then only the elements from this set are involved in the relation.

3 The relation from the set {0, 1, 2} to the set {0, 1, 4} could be *is the square of*.

4 The mapping $x \to \frac{1}{x}$ cannot be defined on the set {0, 1, 2}.

5 If there is a mapping from set A to set B, and set A contains 5 elements, then set B must contain at least 5 elements.

Questions 6–15: On the answer sheet, circle A, or B, or C, or D, or E (don't know).

Questions 6 and 7 refer to Figure 1

6 The diagram which does *not* show a mapping from P to Q is
A (i) ***B*** (ii) ***C*** (iii) ***D*** (iv) ***E***

7 For which diagram is it true to say that 'The relation from P to Q is a mapping, and the relation from Q to P is also a mapping'?
A (i) ***B*** (ii) ***C*** (iii) ***D*** (iv) ***E***

8 A mapping from set X to set Y is expressed by the open sentence 'x exceeds three times y by 1', where $x \in X$ and $y \in Y$. Then
A $x = 3y+1$ ***B*** $x = 3y-1$ ***C*** $x+3y = 1$
D $x+3y = -1$ ***E***

9 The mapping from the set $P = \{1, 2, 3\}$ to the set of rational numbers, given by $x \to \frac{1}{x}$, is

A $\{1, 2, 3\}$ *B* $\{1, \frac{1}{2}, \frac{1}{3}\}$ *C* $\{(1, 1), (2, \frac{1}{2}), (3, \frac{1}{3})\}$
D $\{(1, 1), (2, 2), (3, 3)\}$ *E*

10 The mapping on the set of integers given by $x \to x+2$ will have a graph whose points lie on a straight line passing through the point

A $(0, 0)$ *B* $(-1, 1)$ *C* $(2, 0)$ *D* $(1, 1)$ *E*

11 The graph of a relation on the set $P = \{1, 2, 3\}$ is shown in Figure 2. If $x \in P$ and $y \in P$, then

A $x \geqslant y$ *B* $x = y$ *C* $x < y$ *D* $x \leqslant y$ *E*

12 The graph of a relation from set $X = \{1, 2, 3\}$ to set $Y = \{1, 2, 3\}$ is shown in Figure 3. If $x \in X$ and $y \in Y$, then

A $y = 1$ *B* $x = 1$ *C* $y > x$ *D* $x > y$ *E*

13 Here are four statements: (i) 'Every relation is a mapping.' (ii) 'Every mapping is a relation.' (iii) 'Every one-to-one correspondence is a mapping.' (iv) 'Every mapping is a one-to-one correspondence.' Which two statements are true?

A (i) and (iii) *B* (i) and (iv) *C* (ii) and (iii)
D (ii) and (iv) *E*

14 The number of mappings possible from set $P = \{a, b\}$ to set $Q = \{c, d\}$ is

A 8 *B* 6 *C* 4 *D* 2 *E*

15 There are four teams in a competition. Each has to play each of the other teams at home and away. How many games have to be played altogether?

A 16 *B* 12 *C* 10 *D* 8 *E*

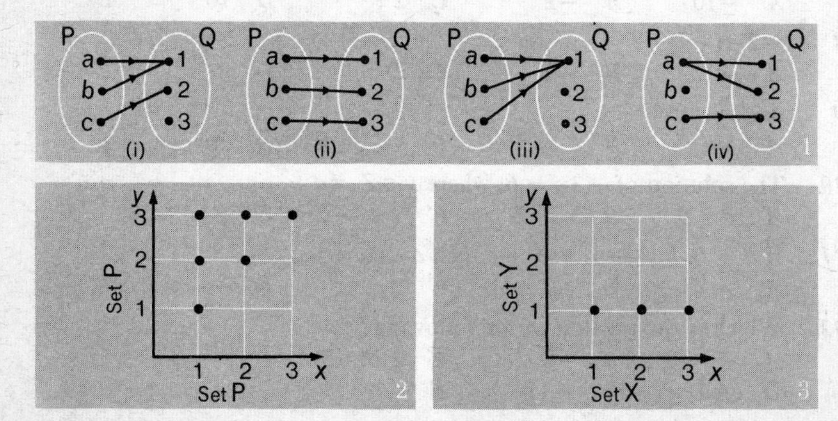

Alg 2

A1

Operations on Integers and Rational Numbers

Questions 1–5: On the answer sheet, circle T for True, F for False, or E for 'don't know'.

---

1 $-13+15 = -2$.

2 $5-(-1) = 6$.

3 $x = -4$ is a solution of $x^2 = 16$.

4 $x^2-7 = -8$ when $x = -1$.

5 $-28 \div -7 = 4$.

Questions 6–15: On the answer sheet, circle A, or B, or C, or D, or E (don't know).

---

6 $3+(-2) =$
A $3-2$ *B* $-3+2$ *C* $-3-2$ *D* $3+2$ *E*

7 $4-(-6) =$
A -10 *B* -2 *C* 2 *D* 10 *E*

8 $-2-(-7)-3 =$
A -12 *B* -2 *C* 2 *D* 4 *E*

9 $a+(a-b) =$
A $-b$ *B* b *C* $-2a+b$ *D* $2a-b$ *E*

10 The solution of $-3x = 6$, where $x \in Z$, is
A $x = 9$ *B* $x = 2$ *C* $x = -2$ *D* $x = -18$ *E*

11 If $a = 0$, $b = -1$, $c = -2$ then $a-bc =$
A -3 *B* -2 *C* 2 *D* 3 *E*

12 Which is the greatest of the following?
A -2×0 *B* -2×1 *C* $2\times(-2)$
D $(-1)\times(-1)\times(-1)$ *E*

13 $-2(-q-r) =$

A $-2q-r$ *B* $2q-r$ *C* $2q+2r$

D $-2q-2r$ *E*

14 $3a-2(a-m) =$

A $5a-2m$ *B* $a+2m$ *C* $a-2m$

D $-6a+2m$ *E*

15 $\frac{24x}{-3x} =$

A $8x$ *B* 8 *C* $-8x$ *D* -8 *E*

Alg 2

A2

Operations on Integers and Rational Numbers

Questions 1–5: On the answer sheet, circle T for True, F for False, or E for 'don't know'.

---

1 $0-(-1) = 1$.

2 $8+(-10) = 2$.

3 $-3\times[(-5)\times 2] = -3\times[5\times(-2)]$.

4 $-1\times(-4)^2 = 16$.

5 $2^2 = (-2)\times(-2)$.

Questions 6–15: On the answer sheet, circle A, or B, or C, or D, or E (don't know).

---

6 $-3+(-3) =$

A -6 *B* 0 *C* 6 *D* 9 *E*

7 $-2-(-6) =$

A -8 *B* -4 *C* 4 *D* 8 *E*

8 Which of the following is false?

A $-2-(-2) = 0$ *B* $6-(-7) = 13$ *C* $8-(-8) = 0$

D $-3-(-4) = 1$ *E*

9 $-5(-2a+b) =$

A $-10a-5b$ *B* $-10a+5b$ *C* $10a+5b$

D $10a-5b$ *E*

10 The solution of $-4x = -20$, where $x \in Z$, is

A $x = 80$ *B* $x = 5$ *C* $x = -5$

D $x = -16$ *E*

11 If $a = -2$, $b = 3$, $x = 0$, $y = -1$, then $ax+by =$

A -3 *B* -1 *C* 1 *D* 3 *E*

12 'In each group of three numbers, the third number is mid-way between the other two on the number line.' For which of the following is this not true?

A $-2, 2, 0$ *B* $-6, 2, -2$ *C* $-5, 6, 1$
D $-1, 5, 2$ *E*

13 $7x-3(x-y) =$

A $4x+y$ *B* $4x-3y$ *C* $4x-y$
D $4x+3y$ *E*

14 $\frac{36ab}{-4ab} =$

A -9 *B* $-9ab$ *C* 9 *D* $9ab$ *E*

15 A square and a rectangle have equal perimeters. The length and breadth of the rectangle are $(3x-3a)$cm and $(x+a)$cm. The length of a side of the square, in cm, is

A $2x-a$ *B* $2x-2a$ *C* $4x-a$
D $4x-2a$ *E*

Alg 2

B1

Operations on Integers and Rational Numbers

Questions 1–5: On the answer sheet, circle T for True, F for false, or E for 'don't know'.

---

1 $(-4)^2 = (-2)^4$.

2 The product of two negative integers is a negative integer.

3 If $x \in Z$ then $-x \in Z$.

4 $-1-(-2) = -1$.

5 $-\frac{8}{10}$, $-\frac{4}{5}$ and $-\frac{12}{16}$ are equivalent fractions.

Questions 6–15: On the answer sheet, circle A, or B, or C, or D, or E (don't know).

---

6 $-6+(-4) =$

A -10 *B* -2 *C* 2 *D* 10 *E*

7 $-3+(-7)-(-2) =$

A -12 *B* -8 *C* 2 *D* 3 *E*

8 Which point does not lie on the line with equation $y = 2x-6$?

A $(1, -4)$ *B* $(0, -6)$ *C* $(-1, -8)$
D $(-3, 0)$ *E*

9 $2x-(y-x) =$

A $x-y$ *B* $-x+y$ *C* $3x-y$ *D* $3x+y$ *E*

10 $(-3)^2-(-2)^2 =$

A -13 *B* -2 *C* 5 *D* 25 *E*

11 The solution set of the equation $(x+1)^2 = 9$, where $x \in Z$, is

A $\{2\}$ *B* $\{2,-4\}$ *C* $\{-2, 4\}$ *D* $\{4\}$ *E*

12 Which replacement for the variable makes the sentence $x^2+5x+10 = 6$ true?

A -4 *B* -2 *C* 0 *D* 2 *E*

13 $-2p(3-2p) =$

A $-5p+4p^2$ *B* $-6p-4p^2$ *C* $-5p-4p^2$

D $-6p+4p^2$ *E*

14 $(4{\cdot}3\times2{\cdot}1)-(4{\cdot}3\times2{\cdot}6) =$

A 20·21 *B* 2·15 *C* −2·15

D −20·21 *E*

15 $\dfrac{4x-8y}{-2} =$

A $2x-4y$ *B* $2x+4y$ *C* $-2x+4y$

D $-2x-4y$ *E*

Alg 2

B2

Operations on Integers and Rational Numbers

Questions 1–5: On the answer sheet, circle T for True, F for False, or E for 'don't know'.

---

1 $-2{\cdot}5$ is to the left of $-5{\cdot}2$ on the number line.

2 If $x \in \{-4, -1, 2\}$ and $y \in \{-2, -1, 0\}$ the greatest value of $x+y$ is 6.

3 $0-(-1) = 1$.

4 $-\frac{4}{5} \div -\frac{1}{2} = \frac{2}{5}$.

5 $(-6 \times -3) \div -2 = -9$.

Questions 6–15: On the answer sheet, circle A, or B, or C, or D, or E (don't know).

---

6 $-1 =$

A $-2+3$ *B* $-3+2$ *C* $-3-2$ *D* $3+2$ *E*

7 If $p \in \{-3, -1, 1\}$ and $q \in \{-4, -2, 0\}$ the greatest and least values of $q-p$ are

A 3, 1 *B* 3, −5 *C* 1, −5 *D* −1, 3 *E*

8 The solution set of the equation $(x-5)^2 = 25$ is

A {5} *B* {0} *C* {−5, 0} *D* {0, 10} *E*

9 $(a-2b+3c)-(3a-2b-c) =$

A $-2a+4b+4c$ *B* $2a-4c$ *C* $-2a-2b-2c$
D $-2a+4c$ *E*

10 $-(-2)^2(-3)^2 =$

A 25 *B* −25 *C* −36 *D* 36 *E*

11 Which replacement for x makes the open sentence $x^2+6x+8 < 0$ true?

A $x = 1$ *B* $x = -1$ *C* $x = -3$
D $x = -5$ *E*

12 Which point does not lie on the line with equation $y = -3x$?

A $(2, -6)$ *B* $(-4, -12)$ *C* $(0, 0)$
D $(-1, 3)$ *E*

13 $5(-2p+q)-7(p-3q) =$

A $-14p+26q$ *B* $-17p+8q$ *C* $-10p+15q$
D $-17p+26q$ *E*

14 $(7{\cdot}1 \times 3{\cdot}8)-(8{\cdot}3 \times 7{\cdot}1) =$

A $-31{\cdot}95$ *B* $-30{\cdot}15$ *C* $30{\cdot}15$
D $31{\cdot}95$ *E*

15 $\dfrac{15a^2-3ab}{3a} =$

A $5a-b$ *B* $5a-3b$ *C* $5a^2-b$
D $5a^2-3b$ *E*

Alg 3

A1

Equations and Inequations in One Variable

(In this paper the variables are on the set of rational numbers, except where stated otherwise.)

Questions 1–5: On the answer sheet, circle T for True, F for False, or E for 'don't know'.

---

1 The negative of 10 is -10.

2 If $p = q$, then $3p = 3q$.

3 $3 \times (-1) > 2 \times (-1)$.

4 $3x = 6$ and $6x = 3$ are equivalent equations.

5 $\{x : -2 < x < 2, x \in Z\} = \{-1, 0, 1\}$.

Questions 6–15: On the answer sheet, circle A, or B, or C, or D, or E (don't know).

---

6 The solution of the equation $x - \frac{3}{4} = 0$ is

A $-\frac{3}{4}$ *B* $\frac{3}{4}$ *C* $-\frac{4}{3}$ *D* $\frac{4}{3}$ *E*

7 The solution of the equation $\dfrac{x}{2} = 7$ is

A -14 *B* $-3\frac{1}{2}$ *C* $3\frac{1}{2}$ *D* 14 *E*

8 The solution of the equation $2y + 5 = 18$ is

A $6\frac{1}{2}$ *B* $11\frac{1}{2}$ *C* 13 *D* 23 *E*

9 The solution set of the equation $3u + 1 = 2u - 6$ is

A $\{-7\}$ *B* $\{-1\}$ *C* $\{\frac{1}{5}\}$ *D* $\{\frac{7}{5}\}$ *E*

10 The sides of a rectangle are $(x + 1)$ cm and $(2x - 3)$ cm long. If the perimeter is 32 cm, the length of the longer side is

A 6 cm *B* 7 cm *C* 8 cm *D* 9 cm *E*

11 The solution set of $x - 3 > 4$ is

A $\{7\}$ *B* $\{1\}$ *C* $\{x : x > 7\}$ *D* $\{x : x > 1\}$ *E*

12 The solution set of $-3y < 12$ is

A $\{y : y > 4\}$ ***B*** $\{y : y > -4\}$ ***C*** $\{y : y < -4\}$
D $\{y : y < 4\}$ ***E***

13 To remove fractions in the equation $\frac{2}{3}x - \frac{1}{4}x = \frac{1}{2}$ you should multiply each side by

A 2 ***B*** 4 ***C*** 6 ***D*** 12 ***E***

14 The sizes of the angles of a triangle are $x°$, $(x+1)°$, $(x+2)°$. $x =$

A 59 ***B*** 61 ***C*** 119 ***D*** 121 ***E***

15 The solution set of $\dfrac{x-6}{3} = 0$ is

A $\{0\}$ ***B*** $\{\frac{1}{2}\}$ ***C*** $\{6\}$ ***D*** $\{9\}$ ***E***

Alg 3

A2

Equations and Inequations in One Variable

(In this paper the variables are on the set of rational numbers, except where stated otherwise.)

Questions 1–5: On the answer sheet, circle T for True, F for False, or E for 'don't know'.

---

1 The reciprocal of $\frac{8}{5}$ is $\frac{5}{8}$.

2 $5\times(-2) > 7\times(-2)$.

3 $\{x : x < 3, x \in Z\} = \{\ldots, -2, -1, 0, 1, 2\}$.

4 If $x > y$ then $-3x > -3y$.

5 If $x = -1$, then $x^2 = x$.

Questions 6–15: On the answer sheet, circle A, or B, or C, or D, or E (don't know).

---

6 The solution of $x+\frac{1}{2} = 0$ is
A 2 *B* $\frac{1}{2}$ *C* $-\frac{1}{2}$ *D* -2 *E*

7 The solution of $4x+9 = 2-3x$ is
A 7 *B* 1 *C* -1 *D* -7 *E*

8 The solution set of $\frac{2}{3}x = 6$ is
A $\{4\}$ *B* $\{5\frac{1}{3}\}$ *C* $\{6\frac{2}{3}\}$ *D* $\{9\}$ *E*

9 The length of a rectangle is $2x$ cm and the breadth is x cm. If the perimeter is more than 60 cm, then
A $x > 5$ *B* $x > 10$ *C* $x > 15$ *D* $x > 20$ *E*

10 The solution set of $x-2 < -1$ is
A $\{x : x < -3\}$ *B* $\{x : x < -1\}$ *C* $\{x : x < 1\}$
D $\{x : x < 3\}$ *E*

11 The solution set of $-2x < -16$ is
A $\{x : x > 8\}$ *B* $\{x : x < 8\}$ *C* $\{x : x > -8\}$
D $\{x : x < -8\}$ *E*

12 In a test a correct answer scores 2, and an incorrect answer scores -1. A boy had x correct scores and 5 incorrect scores. If his total score was 21, $x =$

A 13 *B* 11 *C* 9 *D* 8 *E*

13 The solution of $\frac{3}{4}x - \frac{2}{3}x = 1$ is

A 3 *B* 4 *C* 6 *D* 12 *E*

14 The solution set of $2(x+3) > x+4$ is

A $\{x : x < 2\}$ *B* $\{x : x > 2\}$ *C* $\{x : x < -2\}$
D $\{x : x > -2\}$ *E*

15 The edges of a cuboid have lengths x cm, $(x+2)$ cm and $(x-4)$ cm. If the total length of the edges is 88 cm, $x =$

A 6 *B* 8 *C* 10 *D* 12 *E*

Alg 3

B1

Equations and Inequations in One Variable

(In this paper the variables are on the set of rational numbers.)

Questions 1–5: On the answer sheet, circle T for True, F for False, or E for 'don't know'.

---

1 The reciprocal of -2 is $-\frac{1}{2}$.

2 If $a+c > a+b$, then $c > b$.

3 $-x < 5 \Leftrightarrow x > -5$.

4 If $x > 0$, $x^2 > x^3$.

5 The sum of the lengths of two sides of a triangle is always greater than the length of the third side.

Questions 6–15: On the answer sheet, circle A, or B, or C, or D, or E (don't know).

---

6 The solution of $\frac{2}{7}x = \frac{7}{2}$ is
A $\frac{49}{4}$ *B* 1 *C* 0 *D* $\frac{49}{2}$ *E*

7 The solution of $2(x-1) = 4-x$ is
A 7 *B* 3 *C* 2 *D* 0 *E*

8 The solution set of $3y-7 = y-(3+2y)$ is
A $\{-2\}$ *B* $\{-1\}$ *C* $\{0\}$ *D* $\{1\}$ *E*

9 The solution set of $2x-5 < 5x-2$ is
A $\{x : x > 1\}$ *B* $\{x : x < -1\}$ *C* $\{x : x > -1\}$
D $\{x : x < 1\}$ *E*

10 One article costs 1 penny less than three times the cost of another. The combined cost of the two articles is 35 pence. The dearer one costs
A 23p *B* 26p *C* 29p *D* 37p *E*

11 $\{x : x < 2\}$ is a subset of
A $\{x : x < 0\}$ *B* $\{x : x > 0\}$ *C* $\{x : x \geqslant 2\}$
D $\{x : x \leqslant 2\}$ *E*

12 The solution of $\frac{1}{3}x-\frac{1}{2}x = 1$ is

A -6 *B* -1 *C* 1 *D* 6 *E*

13 If W = {whole numbers}, N = {natural numbers}, Z = {integers}, Q = {rational numbers}, 0, -10 and $\frac{20}{3}$ are all members of the set

A W *B* N *C* Z *D* Q *E*

14 The solution set of $1-\frac{1}{3}(5-x) > 0$ is

A $\{x : x < -2\}$ *B* $\{x : x > -2\}$ *C* $\{x : x > 2\}$
D $\{x : x < 2\}$ *E*

15 The sum of three consecutive even integers, n, $n+2$ and $n+4$, exceeds 420. The least possible value of the middle number is

A 138 *B* 140 *C* 142 *D* 144 *E*

Alg 3

B2

Equations and Inequations in One Variable

(In this paper the variables are on the set of rational numbers.)

Questions 1–5: On the answer sheet, circle T for True, F for False, or E for 'don't know'.

---

1 If each side of the equation $2x = 3$ is multiplied by zero a true sentence is obtained.

2 $-\dfrac{2}{x} = -\dfrac{1}{3} \Leftrightarrow x = 6.$

3 If $a < 0$ and $x > y$ then $ax > ay$.

4 The solution set of $x^2 = 1$ is $\{1, -1\}$.

5 If $x^2 > 1$, then x must be greater than 1.

Questions 6–15: On the answer sheet, circle A, or B, or C, or D, or E (don't know).

---

6 The reciprocal of $-\frac{1}{3}$ is
A 3 *B* 1 *C* $\frac{1}{3}$ *D* -3 *E*

7 The solution of $6x-1 = 4(5-2x)$ is
A $\frac{3}{2}$ *B* 1 *C* $\frac{1}{2}$ *D* 0 *E*

8 The solution set of $3-\frac{1}{2}x > 1$ is
A $\{x : x > -4\}$ *B* $\{x : x < 4\}$ *C* $\{x : x > 8\}$
D $\{x : x < -8\}$ *E*

9 The length of a rectangle is $4x$ m and its breadth is x m. If the area of the rectangle exceeds 144 m², the minimum perimeter exceeds
A 30 m *B* 48 m *C* 60 m *D* 120 m *E*

10 The solution set of $8-2y > y+11$ is
A $\{y : y > -1\}$ *B* $\{y : y > 1\}$ *C* $\{y : y < -1\}$
D $\{y : y < 1\}$ *E*

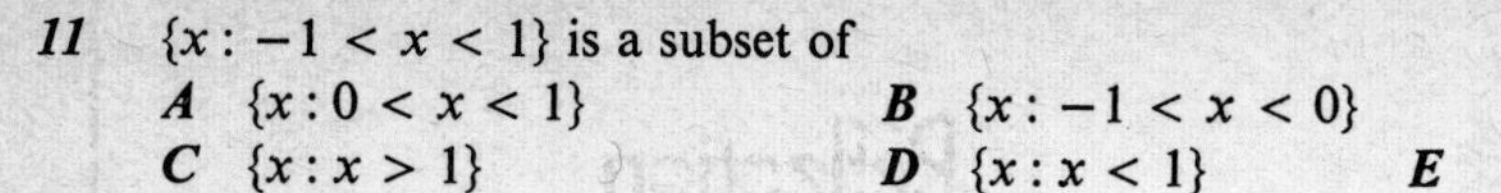

11 $\{x : -1 < x < 1\}$ is a subset of

A $\{x : 0 < x < 1\}$ *B* $\{x : -1 < x < 0\}$
C $\{x : x > 1\}$ *D* $\{x : x < 1\}$ *E*

12 A test contains 60 questions. 3 marks are given for each correct answer, and 2 marks are deducted for each question which is not correctly answered. If the pass mark is 50, the minimum number of correct answers required for a pass mark is

A 40 *B* 34 *C* 28 *D* 22 *E*

13 If $p > q$, which of the following is false?

A $2p > 2q$ *B* $2q < 2p$ *C* $-2p > -2q$
D $-2p < -2q$ *E*

14 The solution of $\frac{x-3}{5} - \frac{x}{2} = 3$ is

A -8 *B* 8 *C* 12 *D* -12 *E*

15 1000 tickets were sold for a concert, some at 75 pence and the rest at 50 pence. If the receipts exceeded £600, the minimum number of dearer tickets sold was

A 401 *B* 501 *C* 601 *D* 701 *E*

Geom 1

A1

Reflection

Questions 1–5: On the answer sheet, circle T for True, F for False, or E for 'don't know'.

Questions 1–5 refer to Figure 1 in which AB is a line of symmetry, Q is the image of P under reflection in AB, and T is the image of S.

1 Under reflection in AB, X maps to X.

2 PQ is perpendicular to AB.

3 PQ is parallel to ST.

4 Angle SXP = angle YXQ.

5 PXQY must be a kite.

Questions 6–15: On the answer sheet, circle A, or B, or C, or D, or E (don't know).

6 The image of (3, 4) in the x-axis is
A (−3, 4) *B* (3, −4) *C* (−3, −4) *D* (3, 4) *E*

7 The number of axes of symmetry of the letter H is
A 1 *B* 2 *C* 3 *D* 4 *E*

8 The image of (−2, 3) in the origin is
A (3, −2) *B* (3, 2) *C* (2, −3) *D* (−2, −3) *E*

9 PQRS is a square, and M is the midpoint of RS. Under reflection in the diagonal PR, the image of M is the midpoint of
A QR *B* PQ *C* PS *D* SR *E*

10 Triangle LMN is isosceles with LM = LN. The triangle fits its outline under reflection in the bisector of
A angle L *B* angle M *C* angle N *E*

11 The image of (0, 2) in the line with equation $x = 2$ is
A(−2, −2) *B* (0, −2) *C* (2, 2) *D* (4, 2) *E*

12 Under reflection in a line PQ, M ↔ M and N ↔ N. The angle between PQ and MN is

A 0° *B* acute *C* 90° *D* obtuse *E*

13 The lengths of the diagonals of a kite are 4 cm and 12 cm. Its area is

A 12 cm² *B* 16 cm² *C* 24 cm² *D* 48 cm² *E*

14 T is a point on a line PQ. The image of A in PQ is B. If angle ATP = 125° then the size of angle BTQ is

A 125° *B* 90° *C* 55° *D* 35° *E*

15 The line from one vertex of a triangle perpendicular to the opposite side is called

A an altitude *B* a median *C* an angle bisector
D a perpendicular bisector *E*

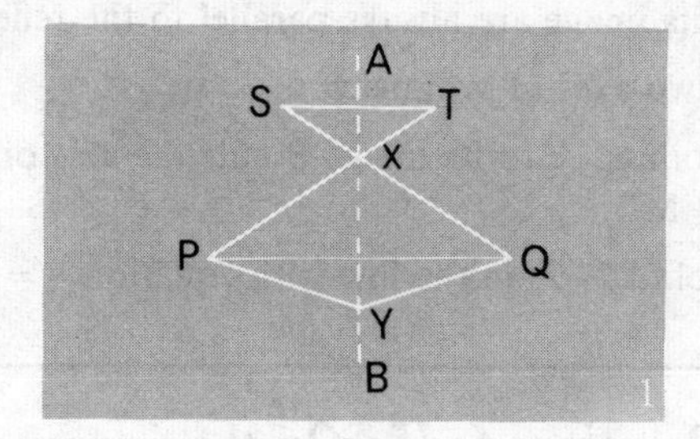

Geom 1

A2

Reflection

Questions 1–5: On the answer sheet, circle T for True, F for False, or E for 'don't know'.

---

1 The image of (0, 4) in the y-axis is (0, −4).

2 A line and its image are always parallel to the reflection axis.

3 A kite has two axes of symmetry.

4 If the line joining A to its image B cuts the axis of reflection at M, then AM = MB.

5 The image of (4, −2) in the line with equation $y = -2$ is (4, −2).

Questions 6–15: On the answer sheet, circle A, or B, or C, or D, or E (don't know).

---

6 Under reflection in the y-axis, $P(3, 5) \leftrightarrow Q(p, q)$ where
A $p = 3, q = 5$ *B* $p = 3, q = -5$ *C* $p = -3, q = 5$
D $p = -3, q = -5$ *E*

7 The number of axes of symmetry of a square is
A 1 *B* 2 *C* 3 *D* 4 *E*

8 One angle of a rhombus is 50°. Two of the other angles are each
A 40° *B* 50° *C* 90° *D* 130° *E*

9 The image of the point (a, b) in the origin is
A $(-a, -b)$ *B* $(-a, b)$ *C* $(-b, a)$ *D* (b, a) *E*

10 Under reflection in the line XY, P ↔ P, Q ↔ R and S ↔ T. Triangle PQS is congruent to triangle
A PST *B* PQR *C* PRT *D* PQT *E*

11 O is the point (0, 0) and K is (4, 4). The image of (4, 0) in OK is
A (0, 4) *B* (0, −4) *C* (−4, 0) *D* (4, 0) *E*

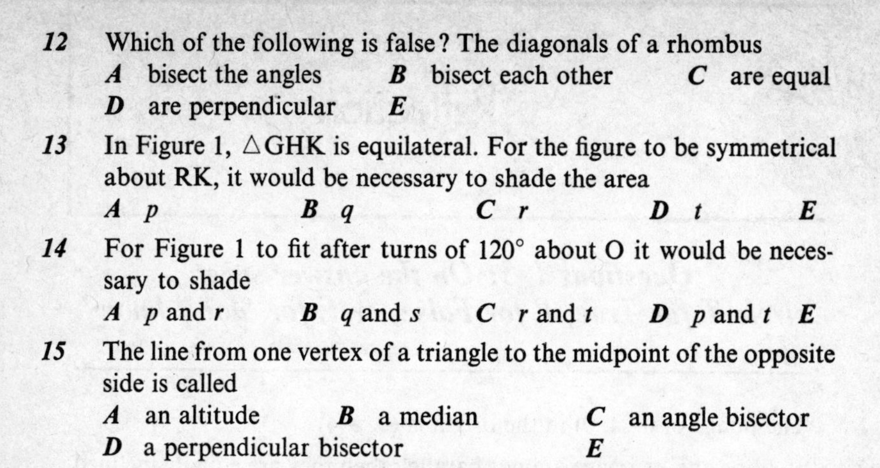

12 Which of the following is false? The diagonals of a rhombus
A bisect the angles *B* bisect each other *C* are equal
D are perpendicular *E*

13 In Figure 1, $\triangle$GHK is equilateral. For the figure to be symmetrical about RK, it would be necessary to shade the area
A p *B* q *C* r *D* t *E*

14 For Figure 1 to fit after turns of 120° about O it would be necessary to shade
A p and r *B* q and s *C* r and t *D* p and t *E*

15 The line from one vertex of a triangle to the midpoint of the opposite side is called
A an altitude *B* a median *C* an angle bisector
D a perpendicular bisector *E*

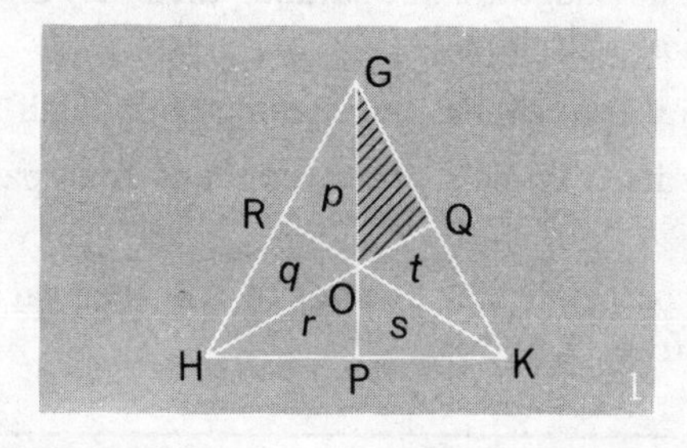

Geom 1

B1

Reflection

Questions 1–5: On the answer sheet, circle T for True, F for False, or E for 'don't know'.

---

1. The image of $(-4, 0)$ in the origin is $(0, -4)$.
2. If a line and its image are not parallel, then they are equally inclined to their axis of reflection.
3. An equilateral triangle has line symmetry *and* half turn symmetry.
4. A kite with its diagonals drawn contains four pairs of congruent triangles.
5. If PQ is the perpendicular bisector of AB, then each point on PQ is equidistant from A and B.

Questions 6–15: On the answer sheet, circle A, or B, or C, or D, or E (don't know).

---

6. P is the point (1, 1), Q is (4, 2) and R is (4, 6). Under reflection in the line of QR, P maps onto S. S is the point
   A $(1, -1)$ *B* $(-1, 1)$ *C* $(7, 1)$ *D* $(4, 1)$ *E*
7. The x-coordinate of the image of (6, 0) in the line with equation $x = a$ is
   A $2a-6$ *B* $6-2a$ *C* $2a+6$ *D* $\frac{1}{2}(2a+6)$ *E*
8. The area in square units of the shape EFGH, where E is (0, 3), F (5, 0), G $(0, -3)$, H (1, 0) is
   A 6 *B* 12 *C* 16 *D* 24 *E*
9. If the size of one angle of a rhombus is $x°$, then one of the other angles is
   A $(180-x)°$ *B* $(360-x)°$ *C* $(90+x)°$ *D* $2x°$ *E*

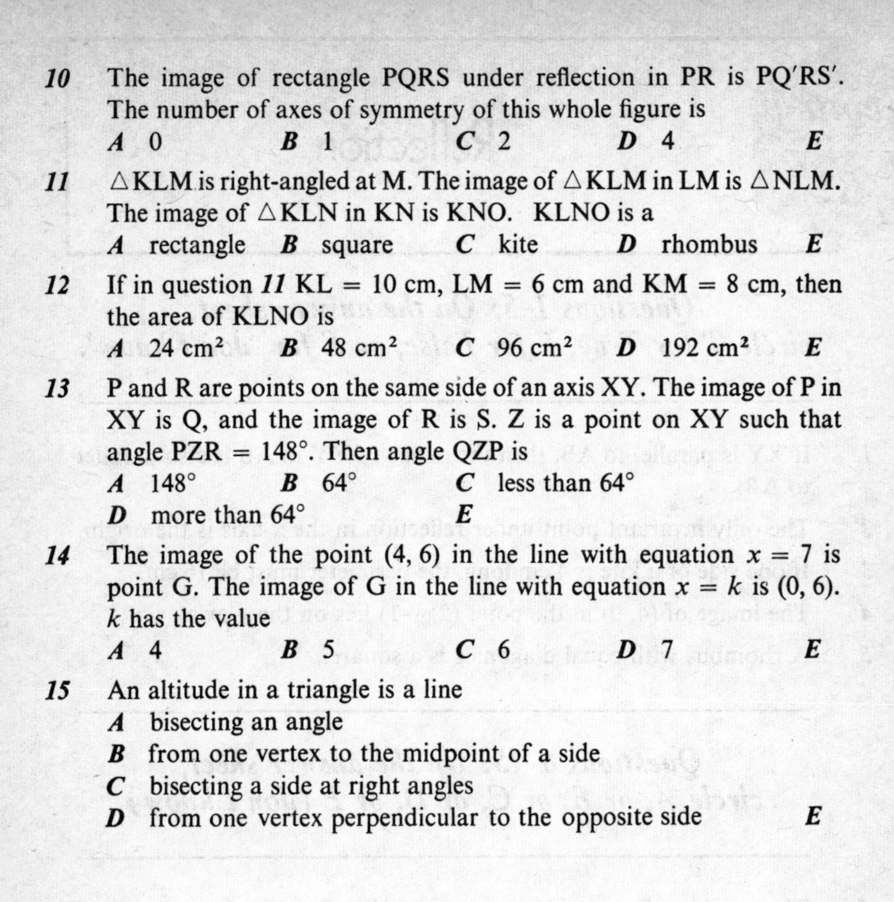

10 The image of rectangle PQRS under reflection in PR is PQ′RS′. The number of axes of symmetry of this whole figure is

A 0 *B* 1 *C* 2 *D* 4 *E*

11 △KLM is right-angled at M. The image of △KLM in LM is △NLM. The image of △KLN in KN is KNO. KLNO is a

A rectangle *B* square *C* kite *D* rhombus *E*

12 If in question *11* KL = 10 cm, LM = 6 cm and KM = 8 cm, then the area of KLNO is

A 24 cm^2 *B* 48 cm^2 *C* 96 cm^2 *D* 192 cm^2 *E*

13 P and R are points on the same side of an axis XY. The image of P in XY is Q, and the image of R is S. Z is a point on XY such that angle PZR = 148°. Then angle QZP is

A 148° *B* 64° *C* less than 64°

D more than 64° *E*

14 The image of the point (4, 6) in the line with equation $x = 7$ is point G. The image of G in the line with equation $x = k$ is (0, 6). k has the value

A 4 *B* 5 *C* 6 *D* 7 *E*

15 An altitude in a triangle is a line

A bisecting an angle

B from one vertex to the midpoint of a side

C bisecting a side at right angles

D from one vertex perpendicular to the opposite side *E*

Geom 1

B2

Reflection

Questions 1–5: On the answer sheet, circle T for True, F for False, or E for 'don't know'.

---

1 If XY is parallel to AB, then the image of XY in AB is also parallel to AB.

2 The only invariant point under reflection in the x-axis is the origin.

3 If one side of a kite is 4 cm long, the perimeter must be 16 cm.

4 The image of (4, 0) in the point (2, −1) lies on the y-axis.

5 A rhombus with equal diagonals is a square.

Questions 6–15: On the answer sheet, circle A, or B, or C, or D, or E (don't know),

---

6 The number of axes of symmetry of a rhombus is
A 1 *B* 2 *C* 3 *D* 4 *E*

7 The image of (3, 0) in the line with equation $x = 3$ is
A (3, 0) *B* (0, 0) *C* (−3, 0) *D* (6, 0) *E*

8 In triangle PQR, PQ = QR = 6 cm, and angle PQR = 90°. Under reflection in PR, Q ↔ S. The area of quadrilateral PQRS is
A 12 cm² *B* 18 cm² *C* 24 cm² *D* 36 cm² *E*

9 A square EFGH has a square EPQR, one of whose angles coincides with angle HEF, cut out from its corner. The remaining shape is symmetrical about
A HF *B* PQ *C* GQ *D* GP *E*

10 O is the origin, and M is the point (6, −6). The image of (4, 1) in OM is
A (−4, −1) *B* (−4, 1) *C* (1, −4) *D* (−1, −4) *E*

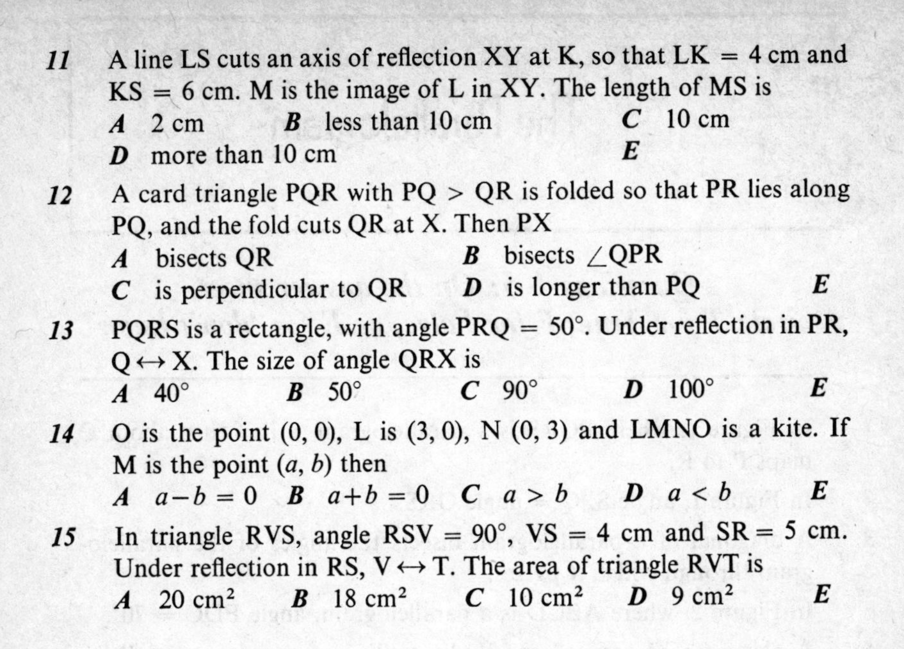

11 A line LS cuts an axis of reflection XY at K, so that LK = 4 cm and KS = 6 cm. M is the image of L in XY. The length of MS is
A 2 cm *B* less than 10 cm *C* 10 cm
D more than 10 cm *E*

12 A card triangle PQR with PQ > QR is folded so that PR lies along PQ, and the fold cuts QR at X. Then PX
A bisects QR *B* bisects ∠QPR
C is perpendicular to QR *D* is longer than PQ *E*

13 PQRS is a rectangle, with angle PRQ = 50°. Under reflection in PR, Q ↔ X. The size of angle QRX is
A 40° *B* 50° *C* 90° *D* 100° *E*

14 O is the point (0, 0), L is (3, 0), N (0, 3) and LMNO is a kite. If M is the point (a, b) then
A $a-b=0$ *B* $a+b=0$ *C* $a>b$ *D* $a<b$ *E*

15 In triangle RVS, angle RSV = 90°, VS = 4 cm and SR = 5 cm. Under reflection in RS, V ↔ T. The area of triangle RVT is
A 20 cm² *B* 18 cm² *C* 10 cm² *D* 9 cm² *E*

The Parallelogram

Questions 1–5: On the answer sheet, circle T for True, F for False, or E for 'don't know'.

---

1 In Figure 1, where PQRS is a parallelogram, a half turn about O maps P to R.

2 In Figure 1, angle SPQ = angle QRS.

3 A diagonal of a parallelogram bisects the angles of the parallelogram through which it passes.

4 In Figure 2, where ABCD is a parallelogram, angle FDC = 70°.

5 A plane can be covered exactly by a tiling of congruent parallelograms.

Questions 6–15: On the answer sheet, circle A, or B, or C, or D, or E (don't know).

---

6 The image of P(1, 3) under reflection in the origin is
A (−1, −3) *B* (−1, 3) *C* (3, 1) *D* (−3, −1) *E*

7 If the parallelogram in Figure 1 is given a half turn about O, then SP maps to
A PQ *B* QR *C* RS *D* SP *E*

8 In every parallelogram the diagonals
A are equal *B* are perpendicular *C* are parallel
D bisect each other *E*

9 P is the point (7, 1), Q is (8, 6) and O is the origin. If OPQR is a parallelogram then the coordinates of R are
A (1, 7) *B* (6, 8) *C* (5, 1) *D* (1, 5) *E*

10 In Figure 3, the angle corresponding to angle 1 is
A 3 *B* 5 *C* 7 *D* 8 *E*

11 In Figure 4, AE is parallel to CB. a is equal to

A 20 *B* 60 *C* 80 *D* 100 *E*

12 In Figure 4, b is equal to

A 20 *B* 60 *C* 80 *D* 100 *E*

13 In Figure 4, c is equal to

A 20 *B* 60 *C* 80 *D* 100 *E*

14 In Figure 5, LMNP is a parallelogram with the dimensions shown. Its area in square cms is

A 10 *B* 20 *C* 30 *D* 40 *E*

15 In Figure 6, PQR is a triangle with sides of different lengths. FGHK is parallel to QR. In the given figure the number of angles equal to angle PGH is

A 1 *B* 2 *C* 3 *D* 4 *E*

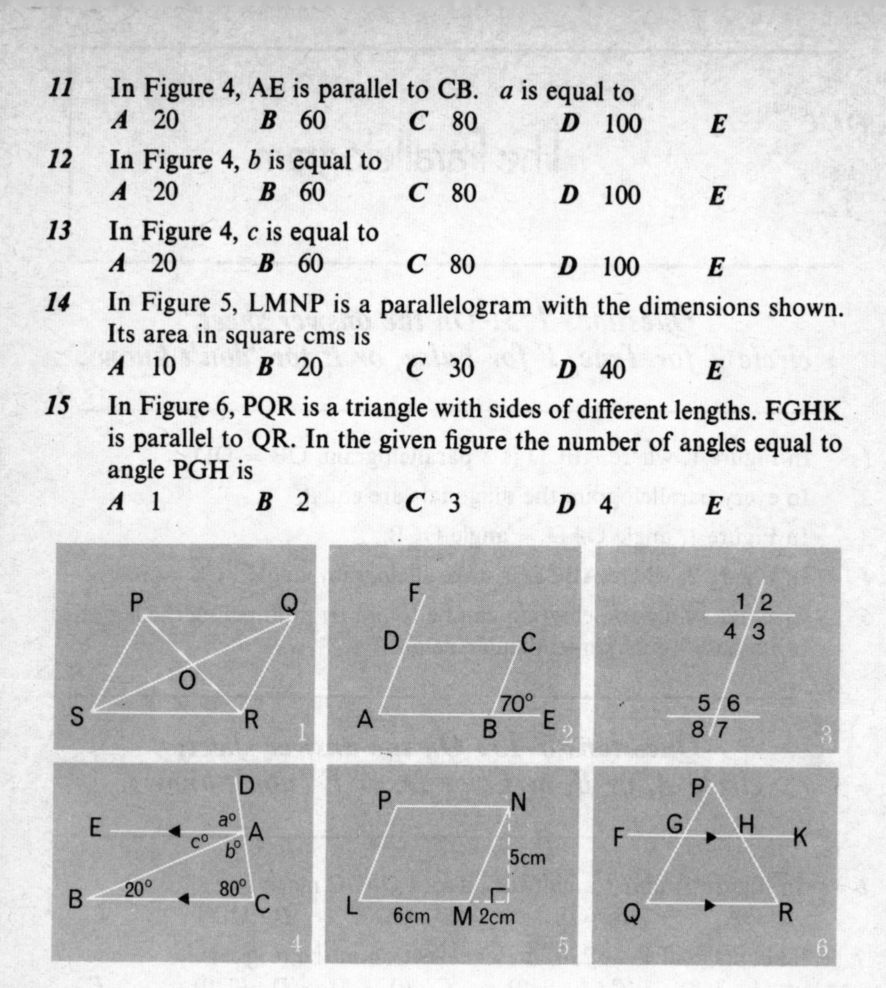

Geom 2

A2

The Parallelogram

Questions 1–5: On the answer sheet, circle T for True, F for False, or E for 'don't know'.

---

1 In Figure 1, where ABCD is a parallelogram, OB = OD.

2 In every parallelogram the diagonals are equal.

3 In Figure 1, angle OAD = angle OCB.

4 In Figure 2, where ABCD is a parallelogram, angle FCB = 70°.

5 The area of a parallelogram can be found by multiplying the length of its base by its perpendicular height.

Questions 6–15: On the answer sheet, circle A, or B, or C, or D, or E (don't know).

---

6 In Figure 1, under a half turn about O, OC maps to
A OA ***B*** OB ***C*** OC ***D*** OD ***E***

7 The image of P (−2, 0) under reflection in the origin is
A (−2, 2) ***B*** (−2, 0) ***C*** (0, −2) ***D*** (2, 0) ***E***

8 In Figure 3, the angle alternate to angle 3 is
A 5 ***B*** 6 ***C*** 7 ***D*** 8 ***E***

9 In Figure 4, ST is parallel to QR. x is equal to
A 110 ***B*** 70 ***C*** 60 ***D*** 50 ***E***

10 In Figure 4, y is equal to
A 110 ***B*** 70 ***C*** 60 ***D*** 50 ***E***

11 In Figure 4, z is equal to
A 110 ***B*** 70 ***C*** 60 ***D*** 50 ***E***

12 A is the point (4, 1), C is the point (1, 3) and O is the origin. If OABC is a parallelogram the coordinates of B are
A (−3, 2) ***B*** (3, −2) ***C*** (4, 5) ***D*** (5, 4) ***E***

13 In Figure 3, the number of angles equal to angle 4 is

A 1 *B* 2 *C* 3 *D* 4 *E*

14 In Figure 5, the area of parallelogram PQRS is 40 cm^2. x is equal to

A 3 *B* 4 *C* 5 *D* 8 *E*

15 A parallelogram has its vertices at the points (0, 0), (6, 0), (8, 3) and (2, 3). Its area in square units is

A 9 *B* 18 *C* 24 *D* 36 *E*

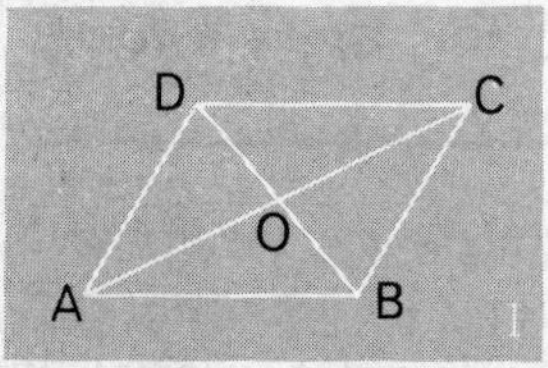

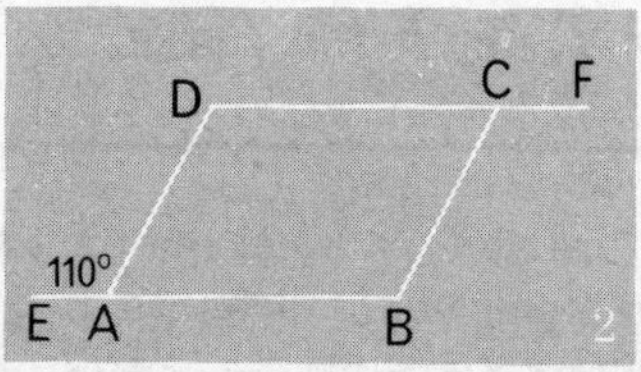

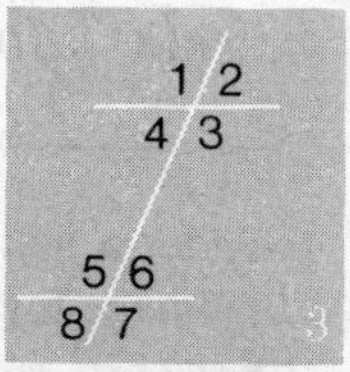

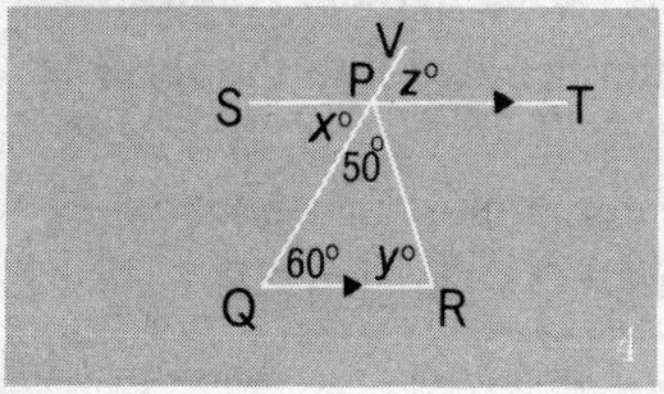

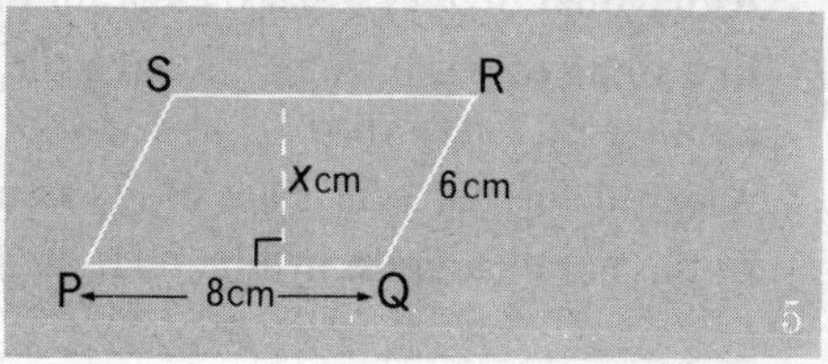

Geom 2

B1

The Parallelogram

Questions 1–5: On the answer sheet, circle T for True, F for False, or E for 'don't know'.

---

1 Under a half turn about a point not on a straight line, the line and its image form a pair of parallel lines.

2 A diagonal divides a parallelogram into two congruent triangles.

3 In a tiling of parallelograms, each tile can be fitted into the space of another by a half turn.

4 A parallelogram which has all its sides equal must be a square.

5 In the parallelogram in Figure 1, angle OPQ = angle OSR.

Questions 6–15: On the answer sheet, circle A, or B, or C, or D, or E (don't know).

---

6 In Figure 1, how many pairs of equal alternate angles are there?
A 2 *B* 4 *C* 6 *D* 8 *E*

7 In Figure 2, DE is parallel to BC. x is equal to
A 35 *B* 65 *C* 80 *D* 115 *E*

8 In Figure 2, y is equal to
A 35 *B* 65 *C* 80 *D* 115 *E*

9 Which of the following must be equal in every parallelogram?
A adjacent sides *B* diagonals *C* opposite angles
D the four angles where the diagonals cross *E*

10 A is the point (2, 3), B is (−3, −3) and C is (4, −3). If ABCD is a parallelogram, then the coordinates of D are
A (9, 3) *B* (3, 9) *C* (−1, −9) *D* (−9, −1) *E*

11 In Figure 3, I, II and III are congruent parallelograms. I can be mapped to III by a half turn about the midpoint of
A PS *B* QR *C* PQ *D* PR *E*

12 The image of the point P (3, 5) under reflection in the point Q (3, 3) is
A (−3, 5) *B* (3, −5) *C* (3, 0) *D* (3, 1) *E*

13 The image of the point P (6, 4) under a half turn is the point P′ (2, 2). The half turn was made about the centre
A (4, 3) *B* (4, 4) *C* (3, 4) *D* (0, 0) *E*

14 In Figure 4, LM and PQ are parallel. $x+y$ is equal to
A 115 *B* 155 *C* 205 *D* 245 *E*

15 The vertices of a parallelogram are the points (0, 0), (5, 3), (10, 0) and (5, −3). Its area in square units is
A 15 *B* 18 *C* 30 *D* 40 *E*

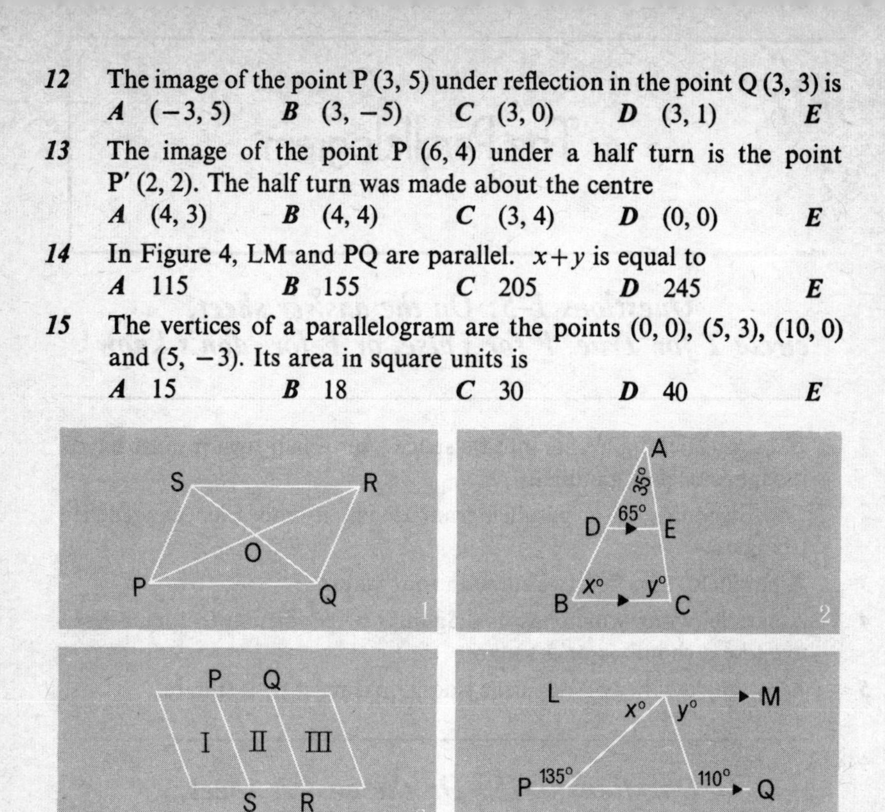

Geom 2

B2

The Parallelogram

Questions 1–5: On the answer sheet, circle T for True, F for False, or E for 'don't know'.

---

1 If a four-sided figure fits into its space after a half turn it must have its opposite sides parallel.

2 The diagonals of a parallelogram divide it into four congruent triangles.

3 A parallelogram fits its outline in four ways.

4 A parallelogram which has its diagonals perpendicular to each other must be a rhombus or a square.

5 A rectangle with two adjacent sides equal must be a square.

Questions 6–15: On the answer sheet, circle A, or B, or C, or D, or E (don't know).

---

6 In Figure 1, PQ is parallel to RS. x is equal to
A 20 *B* 35 *C* 40 *D* 70 *E*

7 In Figure 1, y is equal to
A 20 *B* 35 *C* 40 *D* 70 *E*

8 In Figure 1, z is equal to
A 20 *B* 35 *C* 40 *D* 70 *E*

9 In Figure 2, AB is parallel to CD. p is equal to
A 35 *B* 55 *C* 120 *D* 125 *E*

10 In Figure 3, PQRS is a parallelogram. In which of the following pairs must the angles be equal to each other?
A ∠OPQ and ∠OPS *B* ∠POQ and ∠QOR
C ∠OPQ and ∠ORS *D* ∠QPS and ∠PQR *E*

11 The image of P (1, 5) under reflection in the point Q (3, 3) is
A (5, 1) *B* (5, 5) *C* (−1, −5) *D* (−5, −1) *E*

12 The image of P $(-5, -5)$ under a half turn is P′ (7, 3). The half turn was made about the point

A $(1, -1)$ *B* $(-1, -1)$ *C* $(0, 0)$ *D* $(6, 4)$ *E*

13 Which of the following figures must have a centre of symmetry but need not have an axis of symmetry?

A rectangle *B* rhombus *C* kite *D* parallelogram *E*

14 The vertices of a parallelogram are (0, 2), (5, 3), (5, 7) and (0, 6). Its area in square units is

A 10 *B* 20 *C* 24 *D* 40 *E*

15 In Figure 4, PQRS is a parallelogram with its diagonals intersecting at O, and QT = SV. Which of the following true statements entitles you to say that PTRV is a parallelogram?

A PT is parallel to VR *B* PT = VR

C TR and PV appear to be parallel

D PTRV maps onto itself under a half turn about O *E*

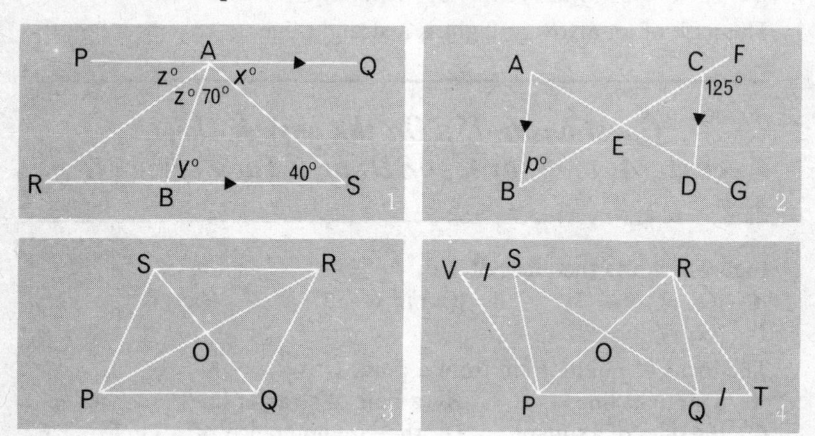

Geom 3

A1

Locus, and Equations of a Straight Line

Questions 1–5: On the answer sheet circle T for True, F for False, or E for 'don't know'.

---

1 The shaded region of Figure 1 is the locus given by $\{P : OP \leqslant 2\}$.

2 The line with equation $y = x$ passes through the origin.

3 The locus given by $\{(x, y) : x = 3\}$ is a line parallel to the x-axis.

4 $\{(x, y) : x = 5\} \cap \{(x, y) : y = 3\} = \{(5, 3)\}$.

5 The path of an arrow in flight is a straight line.

Questions 6–15: On the answer sheet, circle A, or B, or C, or D, or E (don't know).

---

6 Figure 2 shows the locus of
A $\{(x, y) : x = 3\}$ *B* $\{(x\ y) : x \leqslant 3\}$ *C* $\{(x, y) : y < 3\}$
D $\{(x, y) : y = 3\}$ *E*

7 The locus of points 1 cm from a point O is
A a straight line *B* a pair of straight lines
C the sides of a square *D* the circumference of a circle *E*

8 $\{(x, y) : x = -3\} \cap \{(x, y) : y = 1\} =$
A $\{(3, 1)\}$ *B* $\{(-3, 1)\}$ *C* $\{(1, -3)\}$ *D* $\{(1, 3)\}$ *E*

9 $\{(x, y) : y = 4\} \cap \{(x, y) : x = 0\} =$
A $\{(4, 4)\}$ *B* $\{(4, 0)\}$ *C* $\{(0, 4)\}$ *D* $\{(0, 0)\}$ *E*

10 The shaded area of Figure 3 is
A $\{(x, y) : 1 \leqslant y < 3\}$ *B* $\{(x, y) : 1 < y \leqslant 3\}$
C $\{(x, y) : 1 \leqslant x < 3\}$ *D* $\{(x, y) : 1 < x \leqslant 3\}$ *E*

11 The lines with equations $y = 6$, $y = -3$, $x = -1$ and $x = 3$ outline a rectangle with an area containing the following number of square units
A 36 *B* 18 *C* 12 *D* 6 *E*

12 The equation of the x-axis is

A $x = 0$ ***B*** $y = 0$ ***C*** $x = y$ ***D*** $x+y = 0$ ***E***

13 $\{(x, y) : y = x+2\} \cap \{(x, y) : x = 1\} =$

A $\{(1, -2)\}$ ***B*** $\{(1, -1)\}$ ***C*** $\{(1, 2)\}$ ***D*** $\{(1, 3)\}$ ***E***

14 The line with equation $y = 4x-8$ cuts the x-axis at the point

A $(2, 0)$ ***B*** $(-2, 0)$ ***C*** $(0, -8)$ ***D*** $(0, 8)$ ***E***

15 Figure 4 shows the locus of

A $\{(x, y) : y = 2x\}$ ***B*** $\{(x, y) : y \leqslant 2x\}$

C $\{(x, y) : y < 2x\}$ ***D*** $\{(x, y) : y > 2x\}$ ***E***

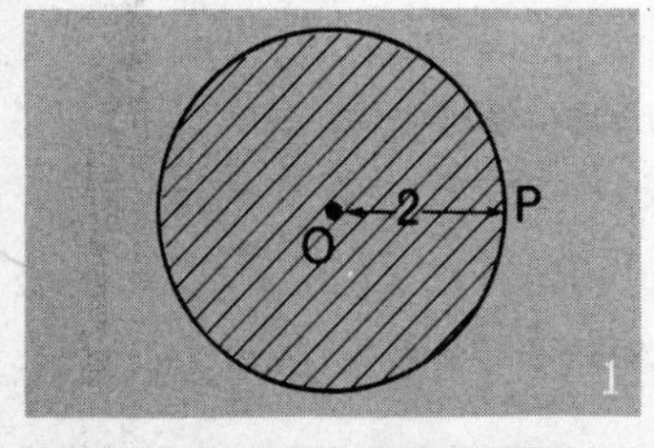

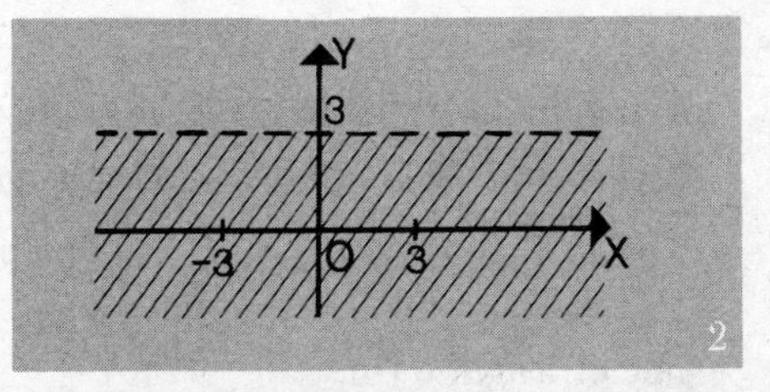

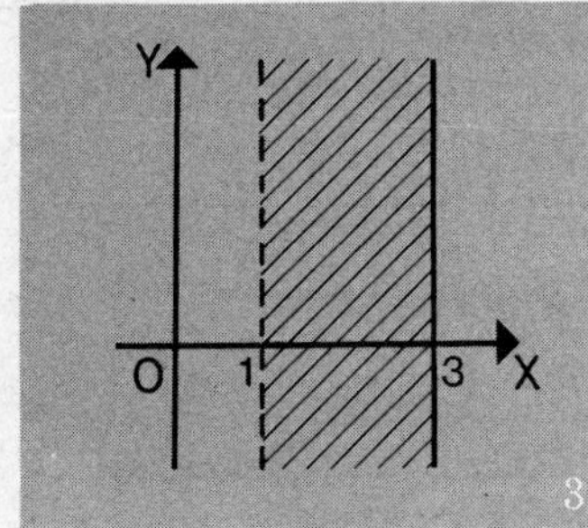

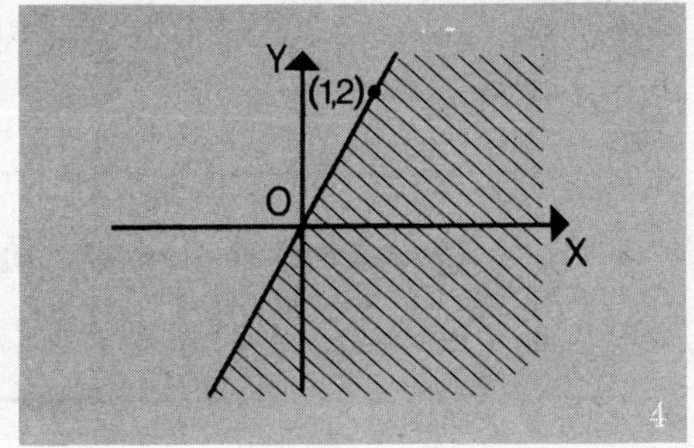

Geom 3

A2

Locus, and Equations of a Straight Line

Questions 1–5: On the answer sheet, circle T for True, F for False, or E for 'don't know'.

---

1 The point (4, 1) lies to the right of the line with equation $x = 2$.

2 The shaded region of Figure 1 is the locus of $\{P : 2 \leqslant OP \leqslant 4\}$.

3 The locus of $\{(x, y) : y = 2\}$ is a line parallel to the x-axis.

4 The locus of a point 2 cm from a line AB of length 6 cm is a single line of length 6 cm.

5 The path of the tip of the hour hand of a clock is the circumference of a circle.

Questions 6–15: On the answer sheet, circle A, or B, or C, or D, or E (don't know).

---

6 The shaded region of Figure 2 shows the locus of
A $\{(x, y) : x < 2\}$ *B* $\{(x, y) : x = 2\}$
C $\{(x, y) : x > 2\}$ *D* $\{(x, y) : x \geqslant 2\}$ *E*

7 $\{(x, y) : x = 5\} \cap \{(x, y) : y = 3\} =$
A $\{(0, 0)\}$ *B* $\{(3, 5)\}$ *C* $\{(5, 0)\}$ *D* $\{(5, 3)\}$ *E*

8 $\{(x, y) : y = 2\} \cap \{(x, y) : y = 4\} =$
A $\{\ \}$ *B* $\{(2, 4)\}$ *C* $\{(4, 2)\}$ *D* $\{(0, 0)\}$ *E*

9 Which of these points lies on the line with equation $y = x - 1$?
A (0, 1) *B* (1, 0) *C* (−1, 0) *D* (−1, −1) *E*

10 The lines with equations $y = -4$, $y = -1$, $x = 2$ and $x = -2$ outline a rectangle with an area containing the following number of square units
A 10 *B* 12 *C* 15 *D* 20 *E*

11 The shaded region of Figure 3 is given by

A $\{(x, y) : -1 < x < 2\}$ *B* $\{(x, y) : -1 \leqslant x \leqslant 2\}$

C $\{(x, y) : -1 < y < 2\}$ *D* $\{(x, y) : -1 \leqslant y \leqslant 2\}$ *E*

12 $\{(x, y) : y = x+1\} \cap \{(x, y) : x = 3\} =$

A $\{(4, 3)\}$ *B* $\{(3, 4)\}$ *C* $\{(3, 2)\}$ *D* $\{(2, 3)\}$ *E*

13 The line with equation $y = -2x+5$ is parallel to the line with equation

A $y = 2x-5$ *B* $y = 2x+5$ *C* $y = -2x-5$ *E*

14 Figure 4 shows the locus of

A $\{(x, y) : y < x+1\}$ *B* $\{(x, y) : y < x-1\}$

C $\{(x, y) : y \leqslant x+1\}$ *D* $\{(x, y) : y \leqslant x-1\}$ *E*

15 All points in the dotted region *R* in Figure 2 bounded by the nearest parts of the *x* and *y*-axes have

A $x < 0$ and $y > 0$ *B* $x < 0$ and $y = 0$

C $x = 0$ and $y = 0$ *D* $x > 0$ and $y > 0$ *E*

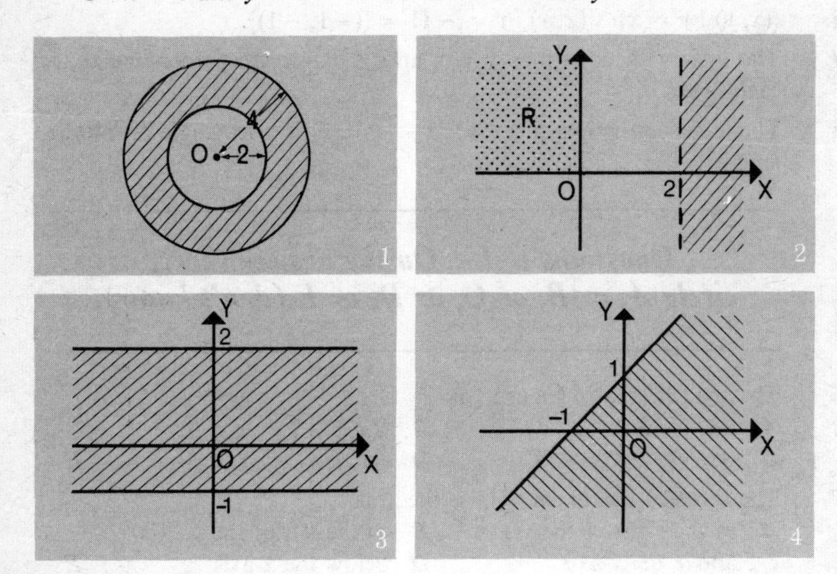

Geom 3

B1

Locus, and Equations of a Straight Line

Questions 1–5: On the answer sheet, circle T for True, F for False, or E for 'don't know'.

---

1 The locus of a point in three dimensions which is 5 cm from a fixed point is the surface of a sphere.

2 The locus given by $\{(x, y) : 0 \leqslant x \leqslant 2,\ -6 \leqslant y \leqslant -2\}$ is the region bounded by a square.

3 $\{(x, y) : y = x\} \cap \{(x, y) : y = -1\} = \{(-1, -1)\}$.

4 The lines with equations $y = x$ and $y = -x$ are perpendicular to each other.

5 The locus consisting of $\{(x, y) : 1 < x < 2\}$ is a rectangle of infinite length.

Questions 6–15: On the answer sheet, circle A, or B, or C, or D, or E (don't know).

---

6 The shaded area in Figure 1 is

A $\{(x, y) : 2 < y < 3\}$ *B* $\{(x, y) : 2 < x < 3\}$
C $\{(x, y) : 2 \leqslant y \leqslant 3\}$ *D* $\{(x, y) : 2 \leqslant x \leqslant 3\}$ *E*

7 The locus of $\{(x, y) : x > 0\}$ is the area

A to the right of the y-axis *B* to the left of the y-axis
C above the x-axis *D* below the x-axis *E*

8 The equation of the line through the origin and the point $(-3, -6)$ is

A $y = -2x$ *B* $y = 2x$ *C* $y = \frac{1}{2}x$ *D* $y = -\frac{1}{2}x$ *E*

9 The shaded region in Figure 2 is the locus of

A $\{(x, y) : x > 0\}$ *B* $\{(x, y) : y > 0\}$
C $\{(x, y) : x > 0\} \cap \{(x, y) : y > 0\}$
D $\{(x, y) : x \geqslant 0\} \cap \{(x, y) : y \geqslant 0\}$ *E*

10 $\{(x, y) : y = x\} \cap \{(x, y) : y = x+2\} =$

A $\{(x, y) : y = x\}$ *B* $\{(x, y) : y = x+2\}$ *C* $\{(0, 2)\}$

D $\varnothing$ *E*

11 The set of points forming the line with equation $y = 3-x$ includes the point

A $(2, -1)$ *B* $(0, -3)$ *C* $(-1, 2)$ *D* $(-1, 4)$ *E*

12 $\{(x, y) : y \geqslant 0\} \cap \{(x, y) : y \leqslant 0\}$ is

A the x-axis *B* the y-axis *C* $\varnothing$ *D* the origin *E*

13 Figure 3 shows the locus of points with equation

A $y = x-2$ *B* $y = \frac{1}{2}x-2$ *C* $y = \frac{1}{2}x+1$

D $y = x+1$ *E*

14 $\{(x, y) : y > x\} \cap \{(x, y) : y = 2\}$ contains the point

A $(-2, -1)$ *B* $(-2, 2)$ *C* $(2, 2)$ *D* $(2, 3)$ *E*

15 Figure 4 shows the intersection of $\{(x, y) : y \leqslant 2-x\}$ and

A $\{(x, y) : y \leqslant -x+1\}$ *B* $\{(x, y) : y \leqslant -x-1\}$

C $\{(x, y) : y \geqslant -x+1\}$ *D* $\{(x, y) : y \geqslant -x-1\}$ *E*

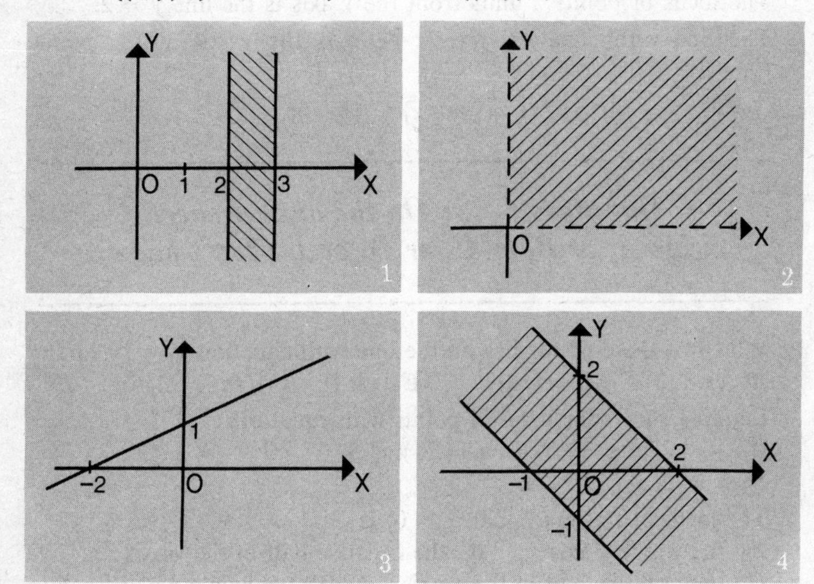

Geom 3

B2

Locus, and Equations of a Straight Line

Questions 1–5: On the answer sheet, circle T for True, F for False, or E for 'don't know'.

---

1 The locus of a point in two dimensions which remains at a constant distance from a fixed point is the region bounded by a circle with centre at the fixed point.

2 The locus $\{(x, y) : -1 \leqslant x \leqslant 1, -2 \leqslant y \leqslant 2\}$ is the region bounded by a square.

3 The locus of points 2 units from the y-axis is the line $y = 2$.

4 The line with equation $y = 3x-2$ cuts the y-axis at the point $(0, -2)$.

5 $\{(x, y) : y = x\} \cap \{(x, y) : y = -x\} = \{(0, 0)\}$.

Questions 6–15: On the answer sheet, circle A, or B, or C, or D, or E (don't know).

---

6 Which of these points lies on the line with equation $y = 5-2x$?
A $(-1, 3)$ ***B*** $(1, 2)$ ***C*** $(2, 1)$ ***D*** $(-2, 5)$ ***E***

7 Figure 1 shows the locus of points with equations
A $y = 2, x \leqslant 3$ ***B*** $x = 2, y \leqslant 3$ ***C*** $x = 2, y < 3$
D $y = 2, x < 3$ ***E***

8 The locus of $\{(x, y) : y < 0, x = 0\}$ is
A the whole y-axis ***B*** the negative half of the y-axis
C the negative half of the x-axis ***D*** the whole x-axis ***E***

9 The locus $\{(x, y) : x < 0, y = -2\}$ contains the point
A $(0, -2)$ ***B*** $(-2, 0)$ ***C*** $(-1, 2)$ ***D*** $(-1, -2)$ ***E***

10 The equation of the line through the origin with gradient $-\frac{1}{3}$ is
A $x+3y = 0$ ***B*** $3x+y = 0$ ***C*** $x-3y = 0$
D $3x-y = 0$ ***E***

11 The heavy black line in Figure 2 is the locus given by

A $\{(x, y) : x < 0\}$ *B* $\{(x, y) : x > 0\}$

C $\{(x, y) : x < 0, y = 0\}$ *D* $\{(x, y) : x = 0, y = 0\}$ *E*

12 $\{(x, y) : x \leqslant 0\} \cap \{(x, y) : x \geqslant 0\}$ is

A the x-axis *B* the y-axis *C* the origin *D* ø *E*

13 The locus given by $\{(x, y) : y = x-2\}$ contains the point

A $(-2, -4)$ *B* $(-2, -5)$ *C* $(-5, -3)$ *D* $(-5, 3)$ *E*

14 Figure 3 shows the line with equation

A $y = -2x+1$ *B* $y = -\frac{1}{2}x+1$ *C* $y = \frac{1}{2}x+1$

D $y = 2x+1$ *E*

15 Figure 4 shows the intersection of $\{(x, y) : y \leqslant -x+2\}$ and

A $\{(x, y) : y \geqslant 0\}$ *B* $\{(x, y) : x \geqslant 0\}$ *C* $\{(x, y) : y \leqslant 2\}$

D $\{(x, y) : x \leqslant 2\}$ *E*

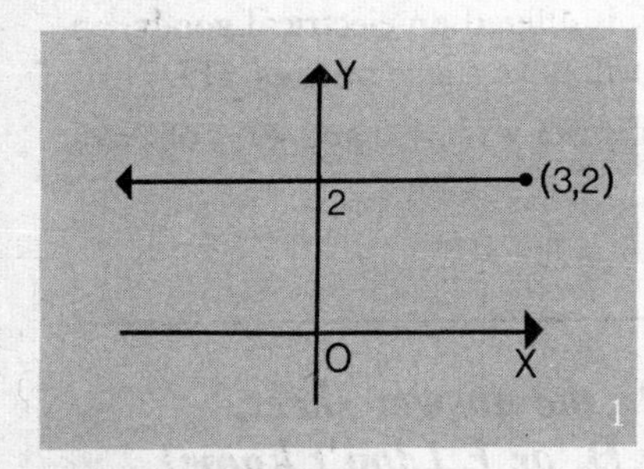

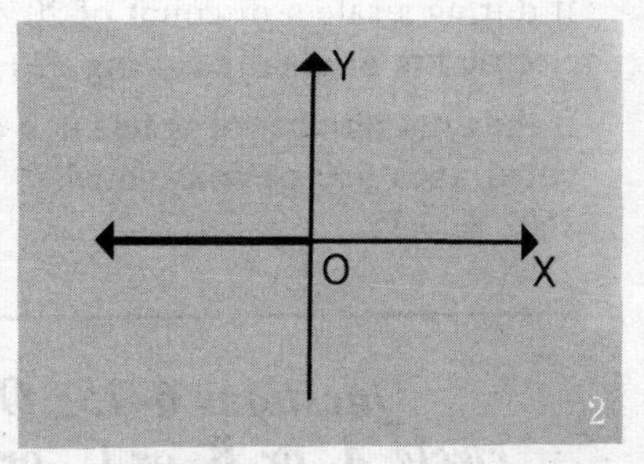

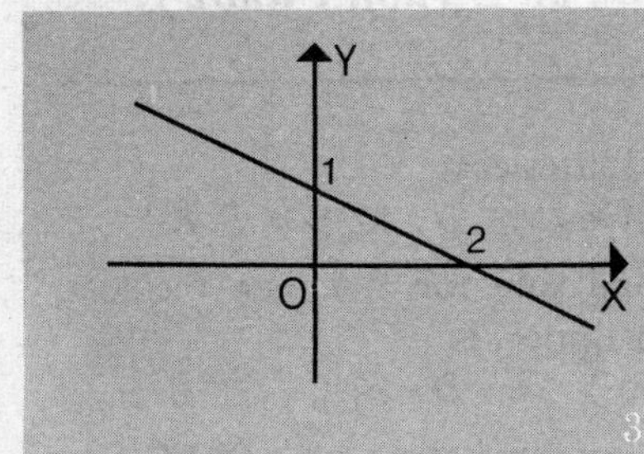

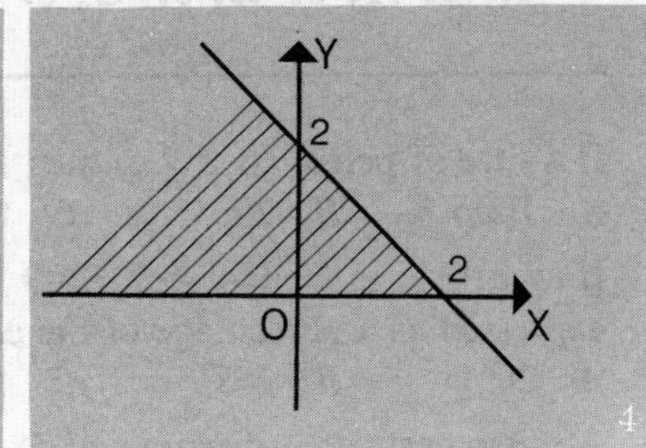

Arith 1

A1

Social Arithmetic

Questions 1–5: On the answer sheet, circle T for True, F for False, or E for 'don't know'.

---

1 If 12 articles cost £1·50 then 120 articles will cost £15.

2 The simple interest in £150 for 1 year at 4% per annum is £6.

3 If during a sale a discount of 20% is offered on electrical goods, an electric fire originally costing £20 will have a sale price of £15.

4 If the total number of voters in a district was 2400 and 40% of them voted, then 960 persons voted.

5 35% = 0·35.

Questions 6–15: On the answer sheet, circle A, or B, or C, or D, or E (don't know).

---

6 If 1 litre of petrol costs $7\frac{1}{2}$p then 12 litres cost
A 100p *B* 90p *C* $84\frac{1}{2}$p *D* $82\frac{1}{2}$p *E*

7 If an article is bought for £100 and sold for £80 then the loss, expressed as a percentage of the cost price, is
A 20% *B* 25% *C* 50% *D* 80% *E*

8 A customer gets a discount of 5% on purchases worth £1000. The discount will be
A £1050 *B* £950 *C* £50 *D* £5 *E*

9 The simple interest on £200 for 6 months at 2% per annum is
A £1 *B* £2 *C* £4 *D* £8 *E*

10 A mixture for concrete consists of 20% cement, 30% stones and the rest sand. In order to make 350 kg of concrete mixture the amount of sand required is
A 35 kg *B* 70 kg *C* 105 kg *D* 175 kg *E*

11 A camera is bought for £12 and is sold so as to make a profit of 25% of the cost price. The selling price will be
A £2·40 *B* £3 *C* £14·40 *D* £15 *E*

12 For a council house gas is charged at a standing charge of £3·75 per quarter plus a charge of 9·5p per therm used. If during a quarter 20 therms are used, the bill for the quarter will be
A £1·81 *B* £1·90 *C* £5·65 *D* £22·75 *E*

13 A garage buys petrol at 7p per litre and sells it at 7½p per litre. In order to make £1 profit they will have to sell
A 20 litres *B* 100 litres *C* 200 litres *D* 1000 litres *E*

14 150% =
A 0·15 *B* 1·5 *C* 15 *D* 150 *E*

15 In the term tests a pupil scored in English 37 out of 50, in French 40 out of 60, in Geography 30 out of 40 and in History 64 out of 80. His highest percentage mark was in
A English *B* French *C* Geography *D* History *E*

Arith 1

A2

Social Arithmetic

Questions 1–5: On the answer sheet, circle T for True, F for False, or E for 'don't know'.

---

1 If a hotel bill comes to £23·70 and 10% is added for 'service', the service charge is £2·37.

2 Amount = Principal + Interest.

3 If in a test a pupil scored 60 out of 120, his percentage mark is 60.

4 225% of £6·40 is £1·60.

5 65% of a certain mixture is sand, so there are 13 kg of sand in 20 kg of the mixture.

Questions 6–15: On the answer sheet, circle A, or B, or C, or D, or E (don't know).

---

6 The simple interest on £100 for 6 months at 3% per anum is

A £6 *B* £3 *C* £1·80 *D* £1·50 *E*

7 If an article is bought for £10 and sold for £5 which of the following is false?

A Loss = £5 *B* Loss = $\frac{1}{2}$ of cost price

C Loss = selling price *D* Loss = cost price *E*

8 The charge for electricity in a house is 2·9p for each of the first 70 units used and 0·8p per unit for all other units used. In a quarter 170 units are used. The bill for the quarter is

A £2·83 *B* £4·09 *C* £14·09 *D* £20·83 *E*

9 5% =

A $\frac{1}{2}$ *B* $\frac{1}{20}$ *C* $\frac{1}{200}$ *D* $\frac{1}{2000}$ *E*

10 An article is bought for £30 and is to be sold so that there will be a profit of 30% of the cost price. The selling price will be

A £20 *B* £21 *C* £39 *D* £40 *E*

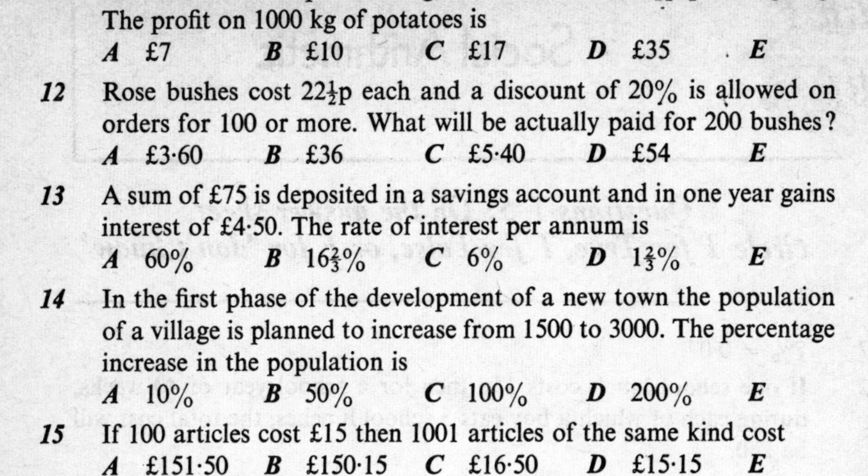

11 Potatoes cost £10 per 1000 kg, and are sold for $8\frac{1}{2}$p per 5 kg bag. The profit on 1000 kg of potatoes is

A £7 *B* £10 *C* £17 *D* £35 *E*

12 Rose bushes cost $22\frac{1}{2}$p each and a discount of 20% is allowed on orders for 100 or more. What will be actually paid for 200 bushes?

A £3·60 *B* £36 *C* £5·40 *D* £54 *E*

13 A sum of £75 is deposited in a savings account and in one year gains interest of £4·50. The rate of interest per annum is

A 60% *B* $16\frac{2}{3}$% *C* 6% *D* $1\frac{2}{3}$% *E*

14 In the first phase of the development of a new town the population of a village is planned to increase from 1500 to 3000. The percentage increase in the population is

A 10% *B* 50% *C* 100% *D* 200% *E*

15 If 100 articles cost £15 then 1001 articles of the same kind cost

A £151·50 *B* £150·15 *C* £16·50 *D* £15·15 *E*

Arith 1

B1

Social Arithmetic

Questions 1–5: On the answer sheet, circle T for True, F for False, or E for 'don't know'.

---

1 $7\% = 0{\cdot}07$.

2 If one school lunch costs 15p then for a school year of 40 weeks, during each of which a boy eats 5 school lunches, the total cost will be £30.

3 If in a test a pupil scores a mark of 35 out of a possible score of 56, his mark is equivalent to $87\frac{1}{2}\%$.

4 1% of £10000 is £100.

5 If the selling price of an article is greater than the cost price, a loss is made.

Questions 6–15: On the answer sheet, circle A, or B, or C, or D, or E (don't know).

---

6 The interest on a loan of £360 for 6 months at 5% per annum is
A £180 *B* £90 *C* £27 *D* £9 *E*

7 Daffodil bulbs are bought at £1·50 per 100 and sold at £2·50 per 100. The profit per cent, reckoned as a percentage of the cost price, is
A $66\frac{2}{3}\%$ *B* 100% *C* $133\frac{1}{3}\%$ *D* 150% *E*

8 An article is sold so as to make a profit of 100% of the cost price. Which of the following is false?
A Cost price = profit *B* Profit = $\frac{1}{2}$ of selling price
C Profit = selling price *D* Cost price = $\frac{1}{2}$ of selling price *E*

9 Gas is charged at 9·5p per therm, plus a standing charge of £3·75 per quarter. The total bill for gas for a quarter in which 120 therms are used will be
A £159 *B* £15·15 *C* £11·40 *D* £4·89 *E*

10 Each side of a square is 10 cm long. The square is changed into a rectangle by increasing the length of two sides by 50% and decreasing the length of the other two sides by 50%. The change in the perimeter of the figure will be

A an increase of 50% *B* an increase of 25%
C a decrease of 25% *D* no change *E*

11 Using the same facts as in question ***10***, the change in the area of the figure will be

A an increase of 50% *B* an increase of 25%
C a decrease of 25% *D* no change *E*

12 During an election, in a certain constituency 18000 persons voted. Of those, 60% voted Conservative and the rest voted Labour. The Conservative candidate won by a majority of

A 1800 votes *B* 3600 votes *C* 7200 votes
D 10800 votes *E*

13 Last year a man earned £30 per week, and of this he saved £5 per week. This year his earnings have increased by 5% and his expenditure has increased by 10%. His weekly savings are now

A £4 *B* £5 *C* £5·25 *D* £5·75 *E*

14 My annual payment for car insurance is quoted at £60 but I am allowed a reduction of 10% because of my association membership. On the sum due after this reduction I am allowed a further reduction of 50% as a no-claims bonus. How much will I actually have to pay?

A £24 *B* £27 *C* £30 *D* £54 *E*

15 I mix 5 kg of a mixture containing 40% of cement and 60% of sand with a further 15 kg of sand. The percentage of cement in the final mixture is

A 40% *B* 25% *C* 20% *D* 10% *E*

Social Arithmetic

Questions 1–5: On the answer sheet, circle T for True, F for False, or E for 'don't know'.

1 If the interest on one Savings Certificate in a year is 5·83p then the interest on a hundred Certificates will be £5·83.

2 During a clearance sale the prices of all goods are reduced by $33\frac{1}{3}\%$, so a coat which originally was priced at £18·60 will be sold for £6·20.

3 If £250 is invested for 6 months at 5% per annum the interest will be £6·25.

4 The postage on one million Christmas cards, each requiring a $2\frac{1}{2}$p stamp, amounts to £25000.

5 If each side of a square of side 1 metre is doubled, the area of the square is increased by 100%.

Questions 6–15: On the answer sheet, circle A, or B, or C, or D, or E (don't know).

6 An article was sold for £30, giving a profit of 100% of the cost price. The cost price was
A £10 *B* £15 *C* £20 *D* £25 *E*

7 35% of a certain mass is 84 kg. The mass is
A 28 kg *B* 175 kg *C* 240 kg *D* 256 kg *E*

8 Discount is allowed at the rate of 1% on the first £10000 of the sales of a factory and at the rate of $1\frac{1}{2}\%$ on the sales over £10000. If during a year a customer buys £20000 worth of goods, his discount will be
A £100 *B* £150 *C* £250 *D* £500 *E*

9 Three-star petrol costs 7p per litre. If only complete litres are sold, the maximum amount of petrol I can buy with £2 is
A 28 litres *B* 29 litres *C* 30 litres *D* 31 litres *E*

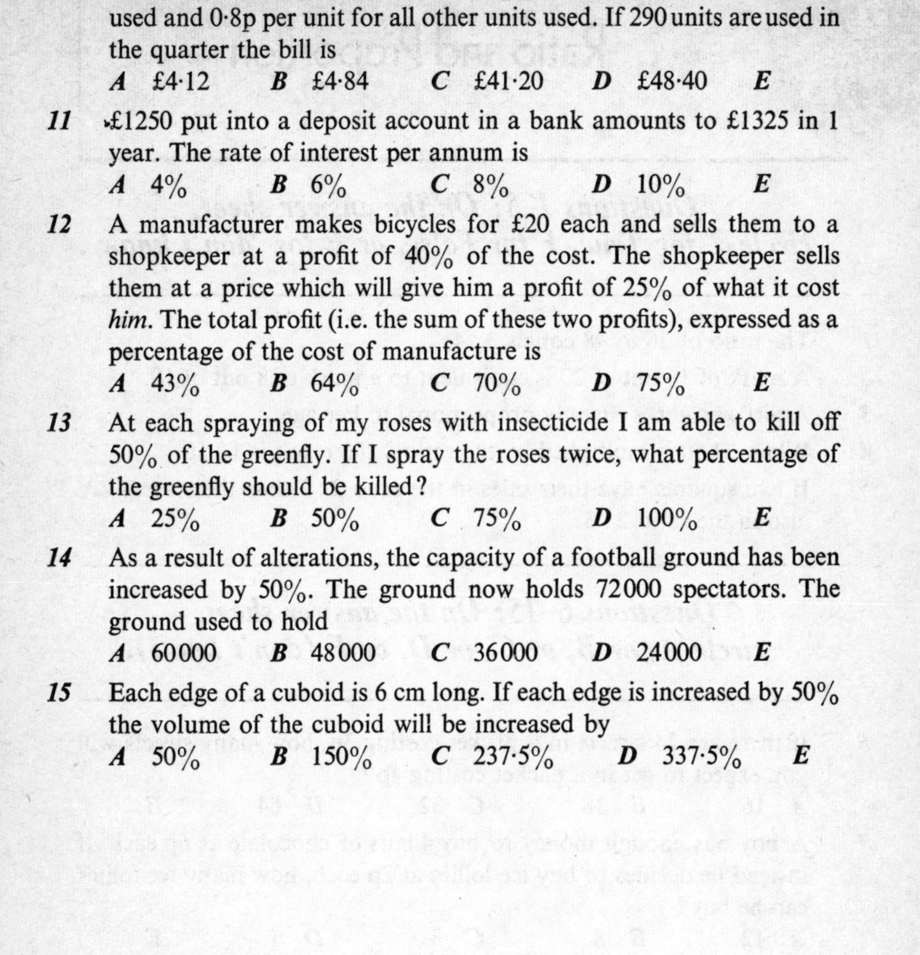

10 Electricity is charged as follows: 2·8p per unit for the first 90 units used and 0·8p per unit for all other units used. If 290 units are used in the quarter the bill is

A £4·12 *B* £4·84 *C* £41·20 *D* £48·40 *E*

11 £1250 put into a deposit account in a bank amounts to £1325 in 1 year. The rate of interest per annum is

A 4% *B* 6% *C* 8% *D* 10% *E*

12 A manufacturer makes bicycles for £20 each and sells them to a shopkeeper at a profit of 40% of the cost. The shopkeeper sells them at a price which will give him a profit of 25% of what it cost *him*. The total profit (i.e. the sum of these two profits), expressed as a percentage of the cost of manufacture is

A 43% *B* 64% *C* 70% *D* 75% *E*

13 At each spraying of my roses with insecticide I am able to kill off 50% of the greenfly. If I spray the roses twice, what percentage of the greenfly should be killed?

A 25% *B* 50% *C* 75% *D* 100% *E*

14 As a result of alterations, the capacity of a football ground has been increased by 50%. The ground now holds 72000 spectators. The ground used to hold

A 60000 *B* 48000 *C* 36000 *D* 24000 *E*

15 Each edge of a cuboid is 6 cm long. If each edge is increased by 50% the volume of the cuboid will be increased by

A 50% *B* 150% *C* 237·5% *D* 337·5% *E*

Arith 2

A1

Ratio and Proportion

Questions 1–5: On the answer sheet, circle T for True, F for False, or E for 'don't know'.

---

1 The ratio of 36 to 48 equals 3 : 4.

2 A mark of 16 out of 20 is equivalent to a mark of 8 out of 10.

3 A girl's height is directly proportional to her age.

4 When £2·80 is multiplied by the ratio $\frac{3}{4}$ the result is £2·10.

5 If two squares have their sides in the ratio 2 : 3 then their areas are also in the ratio 2 : 3.

Questions 6–15: On the answer sheet, circle A, or B, or C, or D, or E (don't know).

---

6 If there are 24 sweets in a packet costing 3p, how many sweets will you expect to get in a packet costing 4p?

A 16 *B* 18 *C* 32 *D* 64 *E*

7 A boy has enough money to buy 4 bars of chocolate at 6p each. If instead he decides to buy ice lollies at 2p each, how many ice lollies can he buy?

A 12 *B* 8 *C* 3 *D* 1 *E*

8 If 6 metres of curtain material cost £4·80, how much will 16 metres of the same material cost?

A £128 *B* £52·80 *C* £30 *D* £12·80 *E*

9 If 6 Belgian francs are worth 5p, the number of Belgian francs I can get for £1 is

A 120 *B* 100 *C* 83 *D* 20 *E*

10 The road between two towns 60 km apart is represented on a map by a line 3 cm long. On the same map the length of the line representing the distance between two towns 80 km apart is

A $2\frac{1}{4}$ cm *B* 4 cm *C* 20 cm *D* $26\frac{2}{3}$ cm *E*

11 A hedge trimmer works at the rate of 3000 cutting strokes per minute. The number of cutting strokes in 2 hours is

A 360000 *B* 180000 *C* 6000 *D* 25 *E*

12 If 50 is multiplied by the factor $\frac{8}{5}$ and the answer is then multiplied by the factor $\frac{7}{2}$, the final number is

A 40 *B* 80 *C* 280 *D* 2800 *E*

13 A school camp has bought enough food to supply 240 pupils with meals for 10 days. How many days will the supplies last if only 200 pupils turn up?

A 8 *B* 9 *C* 10 *D* 12 *E*

14 If the R.F. of a map is 1 : 20000, then 1 cm on the map represents

A 2 m *B* 20 m *C* 200 m *D* 2000 m *E*

15 An alloy is made of copper and tin in the ratio 7 : 3. If in a bar of alloy there are 63 kilogrammes of copper, the number of kilogrammes of tin is

A 21 *B* 27 *C* 100 *D* 147 *E*

Arith 2

A2

Ratio and Proportion

Questions 1–5: On the answer sheet, circle T for True, F for False, or E for 'don't know'.

---

1 The ratio of 30p to £2 is 3 : 20.

2 A score of 60 out of 75 is equivalent to 80 out of 100.

3 The speed of a racing car and the time it takes to complete a race are in inverse proportion.

4 If the side of a cube is doubled then its volume will also be doubled.

5 James the Fourth reigned for 26 years so James the Second must have reigned for 13 years.

Questions 6–15: On the answer sheet, circle A, or B, or C, or D, or E (don't know).

---

6 The line joining two towns 60 km apart is represented by a line 5 cm long on a map. What is the distance between two towns which are shown to be 7 cm apart on the map?

A 35 km *B* $42\frac{6}{7}$ km *C* $77\frac{1}{7}$ km *D* 84 km *E*

7 A boy can put 500 leaflets into envelopes in 40 minutes. Working at the same rate, the number of minutes he takes to put 400 leaflets into envelopes is

A 5 *B* 32 *C* 50 *D* 320 *E*

8 The import duty to be paid on a watch worth £24 is £4·50. At the same rate, what is the duty to be paid on a watch worth £36?

A £3 *B* £6 *C* £6·75 *D* £13·50 *E*

9 Using a mower with blades 32 cm long, a man takes 50 minutes to mow his lawn. Assuming he pushes the mower at the same rate, how many minutes will he take if he uses a mower with blades 40 cm long?

A 10 *B* $12\frac{1}{2}$ *C* 40 *D* $62\frac{1}{2}$ *E*

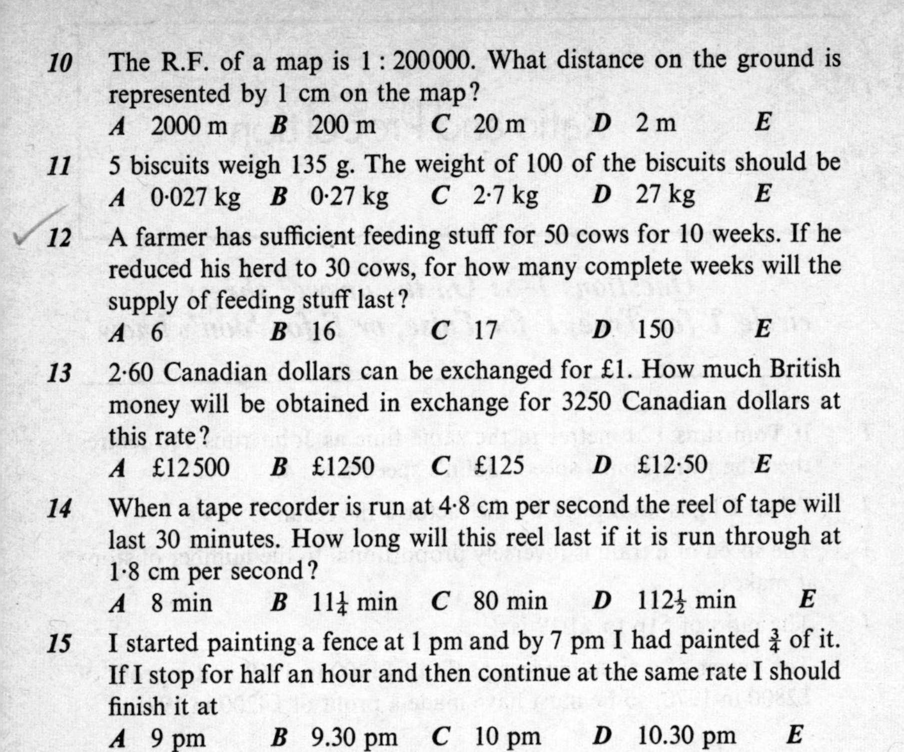

10 The R.F. of a map is 1 : 200000. What distance on the ground is represented by 1 cm on the map?
A 2000 m *B* 200 m *C* 20 m *D* 2 m *E*

11 5 biscuits weigh 135 g. The weight of 100 of the biscuits should be
A 0·027 kg *B* 0·27 kg *C* 2·7 kg *D* 27 kg *E*

12 A farmer has sufficient feeding stuff for 50 cows for 10 weeks. If he reduced his herd to 30 cows, for how many complete weeks will the supply of feeding stuff last?
A 6 *B* 16 *C* 17 *D* 150 *E*

13 2·60 Canadian dollars can be exchanged for £1. How much British money will be obtained in exchange for 3250 Canadian dollars at this rate?
A £12500 *B* £1250 *C* £125 *D* £12·50 *E*

14 When a tape recorder is run at 4·8 cm per second the reel of tape will last 30 minutes. How long will this reel last if it is run through at 1·8 cm per second?
A 8 min *B* $11\frac{1}{4}$ min *C* 80 min *D* $112\frac{1}{2}$ min *E*

15 I started painting a fence at 1 pm and by 7 pm I had painted $\frac{3}{4}$ of it. If I stop for half an hour and then continue at the same rate I should finish it at
A 9 pm *B* 9.30 pm *C* 10 pm *D* 10.30 pm *E*

Arith 2

B1

Ratio and Proportion

Questions 1–5: On the answer sheet, circle T for True, F for False, or E for 'don't know'.

---

1 If Tom runs 120 metres in the same time as John runs 150 metres then the ratio Tom's speed : John's speed is 5 : 4.

2 When 2 kg is multiplied by the factor $\frac{5}{8}$ the result is 125 g.

3 The speed of a train is inversely proportional to the number of stops it makes.

4 The ratio of 51p to £1·19 is $\frac{3}{7}$.

5 The owner of a shop made a profit of £1400 in 1969 and a profit of £2800 in 1970, so he must have made a profit of £4200 in 1971.

Questions 6–15: On the answer sheet, circle A, or B, or C, or D, or E (don't know).

---

6 A girl sets out with enough money to buy 6 tins of dog food costing 15p per tin. She finds that the price has been reduced to 12p per tin. How many tins can she buy with the money?
A 4 *B* 5 *C* 7 *D* 8 *E*

7 Multiplying by the factor $\frac{7}{15}$ and then by the factor $\frac{3}{28}$ is the same as multiplying by the factor
A $\frac{1}{20}$ *B* $\frac{2}{9}$ *C* $\frac{10}{43}$ *D* $\frac{3}{4}$ *E*

8 795 Japanese yen can be exchanged for £1. At the same rate, the number of yen obtained for £4·60 will be
A 36·57 *B* 365·7 *C* 3657 *D* 36570 *E*

9 The R.F. of a map is 1 : 500000, so 1 cm on the map represents
A 5000 m *B* 500 m *C* 50 m *D* 5 m *E*

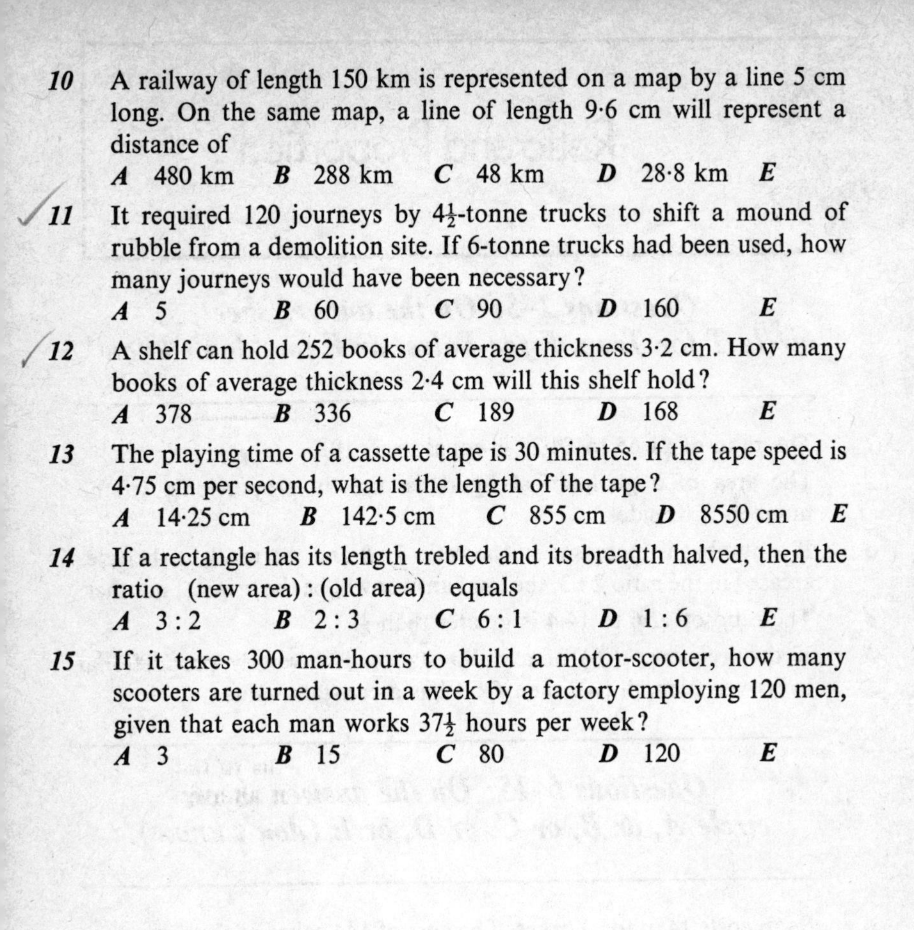

10 A railway of length 150 km is represented on a map by a line 5 cm long. On the same map, a line of length 9·6 cm will represent a distance of

A 480 km *B* 288 km *C* 48 km *D* 28·8 km *E*

11 It required 120 journeys by $4\frac{1}{2}$-tonne trucks to shift a mound of rubble from a demolition site. If 6-tonne trucks had been used, how many journeys would have been necessary?

A 5 *B* 60 *C* 90 *D* 160 *E*

12 A shelf can hold 252 books of average thickness 3·2 cm. How many books of average thickness 2·4 cm will this shelf hold?

A 378 *B* 336 *C* 189 *D* 168 *E*

13 The playing time of a cassette tape is 30 minutes. If the tape speed is 4·75 cm per second, what is the length of the tape?

A 14·25 cm *B* 142·5 cm *C* 855 cm *D* 8550 cm *E*

14 If a rectangle has its length trebled and its breadth halved, then the ratio (new area) : (old area) equals

A 3 : 2 *B* 2 : 3 *C* 6 : 1 *D* 1 : 6 *E*

15 If it takes 300 man-hours to build a motor-scooter, how many scooters are turned out in a week by a factory employing 120 men, given that each man works $37\frac{1}{2}$ hours per week?

A 3 *B* 15 *C* 80 *D* 120 *E*

Arith 2

B2

Ratio and Proportion

Questions 1–5: On the answer sheet, circle T for True, F for False, or E for 'don't know'.

---

1 The ratio of £1·65 to £4·29 is equal to 5 : 19.

2 The area of a geometrical figure is directly proportional to the number of its sides.

3 If a number is increased in the ratio 3 : 2 and the result is then decreased in the ratio 2 : 3, the final answer will be the original number.

4 The ratio of 12·6 to 14·4 is greater than $\frac{3}{4}$.

5 If the frequency (f) of a radio station is inversely proportional to its wave-length (w), then $f \times w$ has a constant value.

Questions 6–15: On the answer sheet, circle A, or B, or C, or D, or E (don't know).

---

6 Soap costs 14½p for 3 cakes. The cost of 144 cakes of the same soap should be

A £7·20 *B* £6·96 *C* £6·67 *D* £6·48 *E*

7 When written by hand an essay averages 10 words to the line and runs to 175 lines. When typed, each line contains an average of 14 words. The number of lines of type will be

A 125 *B* 171 *C* 179 *D* 245 *E*

8 A geometrical drawing is enlarged. A line 8·4 cm long on the original becomes 21·0 cm long on the enlargement. A line 4·8 cm long on the original will have on the enlargement a length of

A 36·75 cm *B* 24·6 cm *C* 17·4 cm *D* 12·0 cm *E*

9 When expressed as the ratio of two numbers in the binary system, the ratio 9 : 17 equals

A $\frac{101}{1001}$ *B* $\frac{101}{10001}$ *C* $\frac{1001}{10001}$ *D* $\frac{1001}{100001}$ *E*

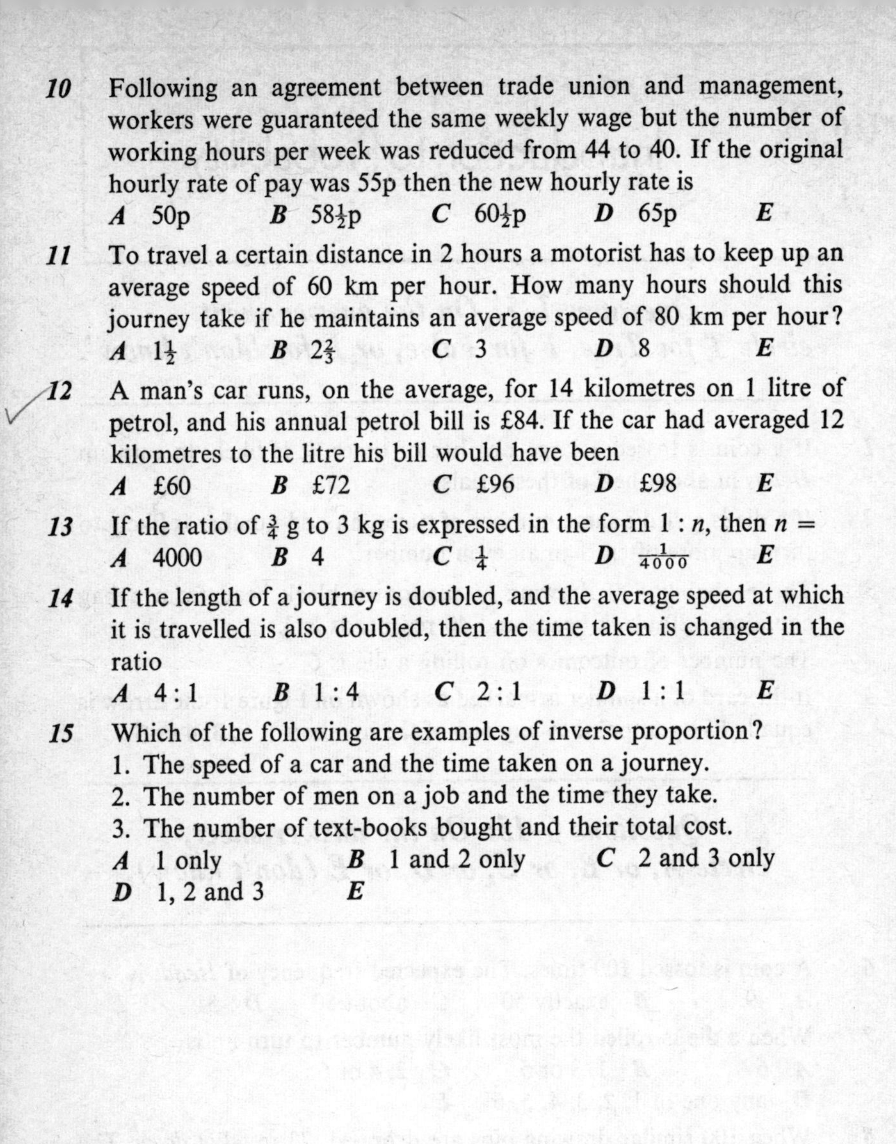

10 Following an agreement between trade union and management, workers were guaranteed the same weekly wage but the number of working hours per week was reduced from 44 to 40. If the original hourly rate of pay was 55p then the new hourly rate is

A 50p *B* $58\frac{1}{2}$p *C* $60\frac{1}{2}$p *D* 65p *E*

11 To travel a certain distance in 2 hours a motorist has to keep up an average speed of 60 km per hour. How many hours should this journey take if he maintains an average speed of 80 km per hour?

A $1\frac{1}{2}$ *B* $2\frac{2}{3}$ *C* 3 *D* 8 *E*

12 A man's car runs, on the average, for 14 kilometres on 1 litre of petrol, and his annual petrol bill is £84. If the car had averaged 12 kilometres to the litre his bill would have been

A £60 *B* £72 *C* £96 *D* £98 *E*

13 If the ratio of $\frac{3}{4}$ g to 3 kg is expressed in the form $1 : n$, then $n =$

A 4000 *B* 4 *C* $\frac{1}{4}$ *D* $\frac{1}{4000}$ *E*

14 If the length of a journey is doubled, and the average speed at which it is travelled is also doubled, then the time taken is changed in the ratio

A 4 : 1 *B* 1 : 4 *C* 2 : 1 *D* 1 : 1 *E*

15 Which of the following are examples of inverse proportion?

1. The speed of a car and the time taken on a journey.
2. The number of men on a job and the time they take.
3. The number of text-books bought and their total cost.

A 1 only *B* 1 and 2 only *C* 2 and 3 only
D 1, 2 and 3 *E*

Arith 3

A

Introduction to Probability

Questions 1–5: On the answer sheet, circle T for True, F for False, or E for 'don't know'.

---

1 If a coin is tossed a large number of times, it is likely to turn up *Heads* in about half of these trials.

2 If a die is rolled a large number of times, an odd number is likely to turn up more often than an even number.

3 The probability of drawing at random a black bead from a bag containing 30 black beads and 40 red beads is $\frac{3}{4}$.

4 The number of outcomes on rolling a die is 6.

5 If the card of a spinner is marked as shown on Figure 1, the arrow is equally likely to point to any one of the numbers 1, 2, 3, 4 or 5.

Questions 6–15: On the answer sheet, circle A, or B, or C, or D, or E (don't know).

---

6 A coin is tossed 100 times. The expected frequency of *Heads* is
A 49 *B* exactly 50 *C* about 50 *D* 51 *E*

7 When a die is rolled the most likely number to turn up is
A 6 *B* 1, 3 or 5 *C* 2, 4 or 6
D any one of 1, 2, 3, 4, 5, 6 *E*

8 When 100 similar drawing pins are dropped, 73 rest *Pin down.* The relative frequency of a pin resting *Pin up* is
A 0·27 *B* 0·73 *C* 27 *D* 73 *E*

9 The probability of a prime number turning up when a die is rolled is
A $\frac{1}{6}$ *B* $\frac{1}{3}$ *C* $\frac{1}{2}$ *D* $\frac{2}{3}$ *E*

10 A class consists of 17 boys and 13 girls. If one name is chosen at random from the class list, the probability that it is a girl's name is
A $\frac{17}{13}$ *B* $\frac{13}{17}$ *C* $\frac{17}{30}$ *D* $\frac{13}{30}$ *E*

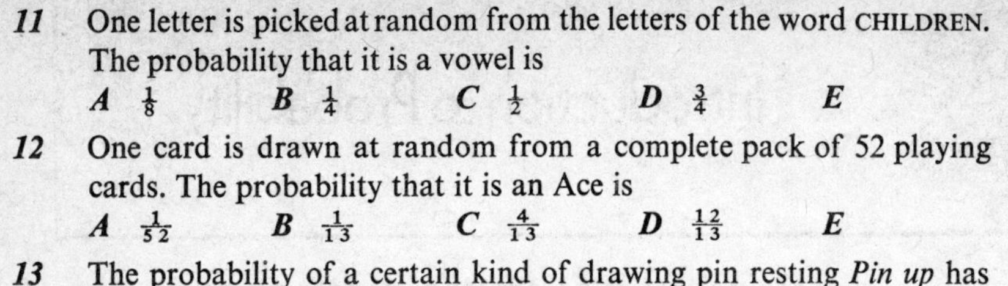

11 One letter is picked at random from the letters of the word CHILDREN. The probability that it is a vowel is

A $\frac{1}{8}$ *B* $\frac{1}{4}$ *C* $\frac{1}{2}$ *D* $\frac{3}{4}$ *E*

12 One card is drawn at random from a complete pack of 52 playing cards. The probability that it is an Ace is

A $\frac{1}{52}$ *B* $\frac{1}{13}$ *C* $\frac{4}{13}$ *D* $\frac{12}{13}$ *E*

13 The probability of a certain kind of drawing pin resting *Pin up* has been found by experiment to be 0·6. If 300 such pins are dropped, the expected frequency of *Pin up* is

A 180 *B* 120 *C* 50 *D* 6 *E*

14 When two dice are rolled together the probability of the total score being 1 is

A 0 *B* $\frac{1}{36}$ *C* $\frac{1}{2}$ *D* 1 *E*

15 When two coins are tossed together, the probability of turning up one *Head* and one *Tail* is

A 0 *B* $\frac{1}{4}$ *C* $\frac{1}{2}$ *D* 1 *E*

Arith 3

B

Introduction to Probability

Questions 1–5: On the answer sheet, circle T for True, F for False, or E for 'don't know'.

---

1 A coin which is tossed 100 times and turns up *Heads* 52 times must be a biased, or unfair, coin.

2 If a die in the shape of a regular tetrahedron (four triangular faces) with the numbers 2, 3, 4 and 5 written on the faces is tossed, the probability that the three numbers visible when it comes to rest are all prime is 1.

3 When a coin is tossed 999 times, the relative frequency of *Heads* turning up in this experiment can never be exactly $\frac{1}{2}$.

4 When two dice are rolled, the probability of a total score of 1 is the same as the probability of a total score of 13.

5 If two coins are tossed, the outcomes may be either *Two heads* or *Two tails* or *One head and one tail*, so the probability of each of these outcomes is $\frac{1}{3}$.

Questions 6–15: On the answer sheet, circle A, or B, or C, or D, or E (don't know).

---

6 A bag contains 15 red beads, 20 blue beads and 25 black beads. One bead is drawn at random. The probability that it is blue is

A $\frac{1}{60}$ ***B*** $\frac{1}{4}$ ***C*** $\frac{1}{3}$ ***D*** $\frac{1}{2}$ ***E***

7 If a whole number from 11 to 20 inclusive is chosen at random, the probability that it will be a prime number is

A $\frac{2}{5}$ ***B*** $\frac{1}{2}$ ***C*** $\frac{3}{5}$ ***D*** $\frac{7}{10}$ ***E***

8 A carton of 10 light bulbs is known to contain 3 faulty bulbs. One is drawn out and found to be faulty and is not replaced. Another bulb is now drawn out. The probability that it too is faulty is

A $\frac{1}{5}$ ***B*** $\frac{2}{9}$ ***C*** $\frac{1}{3}$ ***D*** $\frac{7}{9}$ ***E***

9 A bag contains 10 black buttons and 1 white button. A button is drawn out and is found to be white. It is not replaced and a second button is drawn out. The probability that it is black is

A 0 ***B*** $\frac{9}{10}$ ***C*** $\frac{10}{11}$ ***D*** 1 ***E***

10 The captain of a football team has guessed the toss of the coin correctly in the first 9 matches of the season. What is the probability that he will guess correctly again in the next (the tenth) match?

A less than $\frac{1}{2}$ ***B*** $\frac{1}{2}$ ***C*** more than $\frac{1}{2}$ ***D*** can't tell ***E***

11 A cube of edge 3 cm is painted all over its faces. It is now cut up into small cubes each of edge 1 cm. If one of these small cubes is chosen at random, the probability that it is painted on exactly two sides is

A $\frac{2}{9}$ ***B*** $\frac{8}{27}$ ***C*** $\frac{4}{9}$ ***D*** $\frac{2}{3}$ ***E***

12 In a certain type of car accident, if a motorist does not wear a safety belt the probability that he will be killed is 0·35. Out of 200 motorists involved in such accidents and not wearing safety belts, how many can be expected to escape death?

A 7 ***B*** 13 ***C*** 70 ***D*** 130 ***E***

13 A coin is tossed and a die is rolled. The probability of getting a *Head* and a *Prime number* is

A $\frac{1}{6}$ ***B*** $\frac{1}{4}$ ***C*** $\frac{1}{3}$ ***D*** $\frac{1}{2}$ ***E***

14 In 1000 trials of an experiment in which 3 coins are tossed at the same time, what is the expected frequency of getting three heads?

A 125 ***B*** 250 ***C*** 333 ***D*** 500 ***E***

15 When two dice are rolled the most likely total score to obtain is

A 5 ***B*** 6 ***C*** 7 ***D*** 8 ***E***

Arith 4
A

Time, Distance, Speed

Questions 1–5: On the answer sheet, circle T for True, F for False, or E for 'don't know'.

Questions 1 and 2 refer to Table 1

1 On the London–Moscow journey, it will be dark when you arrive in Poznan.

2 It takes 6 h 5 min to travel from the Hook of Holland to Hanover.

3 In Figure 1 the average speed of the slow train is about 70 km/h.

4 Walking at 6 km/h, it will take more than 7 hours to travel 45 km.

5 An object moving at an average speed of 10 cm/s will travel 5 cm in 2 seconds.

Questions 6–15: On the answer sheet, circle A, or B, or C, or D, or E (don't know).

Questions 6–8 refer to Table 1

6 If I leave London on Monday, I arrive in Moscow on
A Monday *B* Tuesday *C* Wednesday *D* Thursday *E*

7 The time spent on the train in Brest is
A 3 h 7 min *B* 9 h 7 min *C* 9 h 47 min
D 20 h 53 min *E*

8 The average speed of the boat between Harwich and the Hook of Holland, to the nearest km/h, is
A 24 *B* 30 *C* 43 *D* 48 *E*

9 In Figure 1, the express train reaches London at
A 2002 *B* 2020 *C* 2024 *D* 2040 *E*

10 In Figure 1 the trains pass one another at a distance from London of
A $302\frac{1}{2}$ km *B* $320\frac{1}{2}$ km *C* 325 km *D* 350 km *E*

11 An object travels 112 cm in 14 seconds. Its average speed is
A $\frac{1}{98}$ cm/s *B* $\frac{1}{8}$ cm/s *C* 8 cm/s *D* 98 cm/s *E*

12 Travelling at an average speed of 7500 km/h a space-craft will travel the 375000 km to the moon in

A 5 hours *B* 50 hours *C* 200 hours *D* 500 hours *E*

13 In 0·5 second an object travelling at 0·4 cm/s will travel

A 0·2 cm *B* 0·8 cm *C* 2 cm *D* 8 cm *E*

14 A boy cycles at a steady speed of 15 km/h. In 25 minutes he will travel

A 4·5 km *B* 6·25 km *C* 40 km *D* 375 km *E*

15 A motorist drives from Hamburg to Nuremberg, 500 km away. By 1400 he has travelled 375 km of the journey when he is delayed by a puncture for $\frac{1}{2}$ hour. What average speed must he keep up for the rest of the journey in order to arrive at 1545?

A $71\frac{3}{7}$ km/h *B* 74 km/h *C* 100 km/h

D $156\frac{1}{4}$ km/h *E*

Table 1. London–Moscow Sealink (Train and Boat) Summer Service

Distance from London in km			*Time*
0	London	dep.	1020
110	Harwich	arr.	1138
		dep.	1200 } *Boat*
300	Hook of Holland	arr.	1815 } *Boat*
		dep.	1900
744	Hanover	arr.	0105
		dep.	0116
1060	Berlin	arr.	0642
		dep.	0710
1320	Poznan	arr.	1159
		dep.	1215
1616	Warsaw	arr.	1612
		dep.	1750
1840	Brest	arr.	2354 M
		dep.	0301 M
2512	Smolensk	arr.	1026 M
2928	Moscow	arr.	1600 M

M: Moscow time. 2 hours must be deducted to give British Summer Time. All other times are the same as British Summer Time.

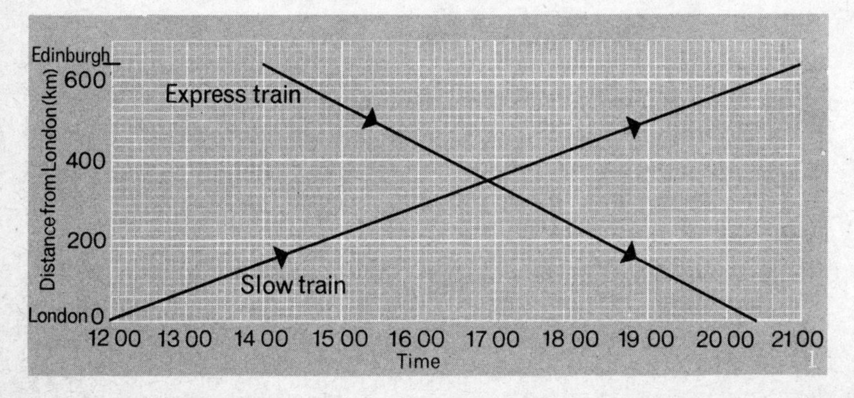

Arith 4 B

Time, Distance, Speed

Questions 1–5: On the answer sheet, circle T for True, F for False, or E for 'don't know'.

---

1 If I leave London for Moscow at 1020 on Wednesday, I can arrange to see a friend in Warsaw station at 5 pm on Thursday (see Table 1).

2 In Figure 1 of Paper A the average speed of the express train is nearly 100 km/h.

3 A B.O.A.C. flight leaves New York at 2030 local time and takes 7 h 50 min for the journey to London. As you have to put your watch forward 5 hours on the journey you will arive at 2220 local time.

4 Light from the sun travelling at 300000 km/s will reach the earth, 150 million km away, in a little more than 8 minutes.

5 I have to catch a boat at Dover at 0300. If I can drive at an average speed of 50 km/h, I must leave Glasgow which is 800 km away, before 1100 the previous day.

Questions 6–15: On the answer sheet, circle A, or B, or C, or D, or E (don't know).

Questions 6–8 refer to Table 1

6 The distance in kilometres from Hanover to Smolensk is
A 2184 *B* 1878 *C* 1768 *D* 1096 *E*

7 The time taken on the train from the Hook of Holland to Berlin is
A 7 h 42 min *B* 11 h 42 min *C* 12 h 18 min
D 12 h 58 min *E*

8 The average speed of the train between Brest and Smolensk, to the nearest km/h, is
A 25 *B* 38 *C* 91 *D* 120 *E*

9 In Figure 1 of Paper A the distance between the two trains at 1524 is
A 230 km *B* 260 km *C* 280 km *D* 298 km *E*

10 The distance travelled in 12 seconds by an object moving at an average speed of 1·5 cm/s is
A 8 cm *B* 18 cm *C* 80 cm *D* 180 cm *E*

11 The time taken to travel 4 cm at an average speed of 0·25 cm/s is
A 1 second *B* 8 seconds *C* 16 seconds
D 160 seconds *E*

12 The average speed of an object which travels 120 cm in 15 seconds, is
A 8 cm/s *B* 12·5 cm/s *C* 18 cm/s *D* 80 cm/s *E*

13 When the slow train in the graph in Figure 1 of Paper A is 100 km nearer Edinburgh than the express train, the time is
A 1600 *B* 1620 *C* 1730 *D* 1800 *E*

14 On a journey a motorist travels for the first 6 hours at an average speed of 80 km/h and then for the next 3 hours at an average speed of 110 km/h. What is his average speed for the whole 9 hour journey?
A 53 km/h *B* 90 km/h *C* 95 km/h *D* 190 km/h *E*

15 The distance from Glasgow to Edinburgh is 60 km. If I keep up an average speed of 60 km/h going from Glasgow to Edinburgh, what average speed must I maintain on the return journey to give an average speed of 120 km/h for the whole journey, there and back?
A 90 km/h *B* 120 km/h *C* 180 km/h
D No such speed is possible *E*

Table 1. London–Moscow Sealink (Train and Boat) Summer Service

Distance from London in km			*Time*	
0	London	dep.	1020	
110	Harwich	arr.	1138	
		dep.	1200	*Boat*
300	Hook of Holland	arr.	1815	*Boat*
		dep.	1900	
744	Hanover	arr.	0105	
		dep.	0116	
1060	Berlin	arr.	0642	
		dep.	0710	
1320	Poznan	arr.	1159	
		dep.	1215	
1616	Warsaw	arr.	1612	
		dep.	1750	
1840	Brest	arr.	2354 M	
		dep.	0301 M	
2512	Smolensk	arr.	1026 M	
2928	Moscow	arr.	1600 M	

M: Moscow time. 2 hours must be deducted to give British Summer Time.
All other times are the same as British Summer Time.